Understanding IC Operational Amplifiers

by Roger Melen and Harry Garland

Howard W. Sams & Co., Inc.
4300 WEST 62ND ST. INDIANAPOLIS, INDIANA 46268 USA

SECOND EDITION
FIRST PRINTING—1978

International Standard Book Number: 0-672-21511-X
Library of Congress Catalog Card Number: 77-99109

Printed in the United States of America.

Preface

Integrated-circuit technology has made possible the low-cost, integrated-circuit operational amplifier—"IC op amp." This book explains how IC op amps work and how they can be used in many practical circuits.

In Chapter 1, the ideal op amp is described. The following three chapters discuss the origin and implications of the non-ideal characteristics found in IC op amps: Chapter 2 reviews basic semiconductor electronics; Chapter 3 explains how integrated op-amp circuitry works; and Chapter 4 discusses practical design considerations in circuits using IC op amps. In these chapters, terms such as *input offset current, output offset voltage, common-mode rejection, slew rate,* and *latch-up* are clearly defined. Many practical circuits are discussed in Chapters 5 and 6, and all component values are given. The final chapter demonstrates how circuits using IC op amps can be interconnected to form complete electronic systems.

In this book, the explanation of how IC op amps work is developed from the fundamental concepts of semiconductor electronics. Once this information is clearly understood, these versatile devices can be used in many new and exciting applications.

This second edition includes descriptions of operational amplifiers that use the latest integrated circuit technologies. These include CMOS, BIMOS, and BIFET op amps. Low-cost IC op amps now offer higher performance than ever before, and this has opened many new areas of application.

In the first edition of this book we had the pleasure of thanking Jim McVittie, Chris Riesbeck, John Linn, Yasuto Miyazawa, Sue Hodge, and Margaret Hogan for their advice and assistance in preparing the manuscript. The influence these people had on the first edition extends to this second edition, though we note that Margaret Hogan is now Margaret Garland.

ROGER MELEN
HARRY GARLAND

Contents

CHAPTER 5

CHAPTER 6

CHAPTER 7

APPENDIX

CHAPTER 1

The Ideal Op Amp

Early operational amplifiers were heavy, costly, fragile tube amplifiers used almost exclusively in analog computers. Today an entire op amp can be fabricated on a tiny chip of silicon for less than a dollar, and these new integrated-circuit op amps have found hundreds of new uses.

In this first chapter we will consider the ideal op amp, since in many circuits IC op amps do behave ideally. Later we will consider the nonideal idiosyncrasies of IC op amps and learn how to work with these characteristics in practical circuit design.

OP-AMP BASICS

An op amp is a very high gain dc amplifier. IC op amps typically have voltage gains in the range of 20,000 to 1,000,000. The schematic symbol of an op amp is shown in Fig. 1-1A. For convenience, the power-supply connections and the ground connection are often not shown, resulting in the simplified symbol of Fig. 1-1B.

As seen in Fig. 1-1, the −input of the op amp is called the *inverting input,* and the +input is called the *noninverting* input. If an input signal is applied to the −input, with the +input grounded, the polarity of the output signal will be opposite to that of the input signal. If an input signal is applied to the +input, with the −input grounded, the polarity of the output signal will be the same as that of the input signal. For an ac signal, this means that the output of the op amp will be 180°

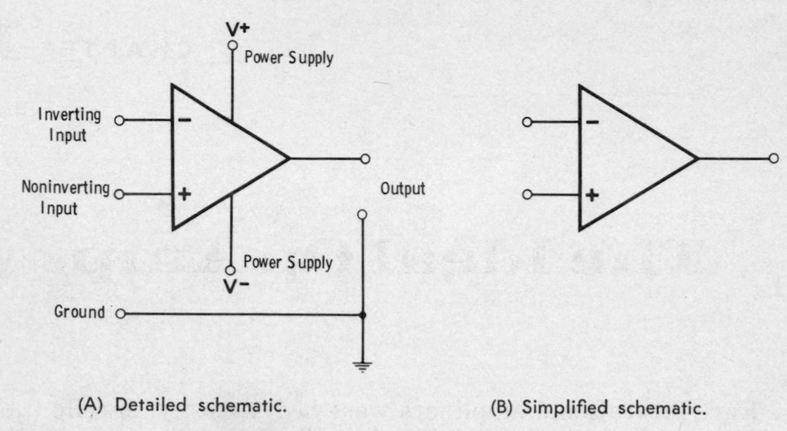

(A) Detailed schematic.

(B) Simplified schematic.

Fig. 1-1. Op-amp schematic symbols.

out of phase with a signal applied to the −input, but in phase with a signal applied to the +input.

If the same signal is applied to both the +input and the −input of the op amp, the two amplified output signals will be 180° out of phase and will completely cancel each other. Since the op amp responds only to differences between its two inputs, it is said to be a *differential amplifier*. The voltage difference between the +input and the −input is called the *differential input voltage*. Since a differential amplifier amplifies only the differential input voltage and is unaffected by signals common to both inputs, it is said to have *common-mode rejection*. Common-mode rejection can be very useful, for example, when measuring small signals in the presence of 60-Hz noise. The 60-Hz noise common to both inputs is rejected, and the op amp amplifies only the small signal difference between the two inputs.

IDEAL OP AMPS WITH NEGATIVE FEEDBACK

The most common op-amp circuit configuration uses two external components: (1) an input component and (2) a feedback component (Fig. 1-2). When the feedback component is between the op-amp output and the −input, the circuit is said to have *negative feedback*. When the feedback component is

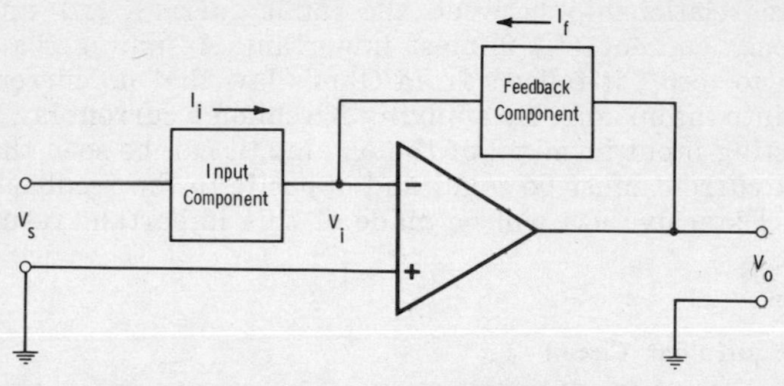

Fig. 1-2. Op-amp circuit with negative feedback.

between the op-amp output and the +input, the circuit is said to have *positive feedback.*

In Fig. 1-2, an op amp is shown in a circuit with negative feedback. V_s is the input signal, V_i is the differential input to the op amp, and V_o is the op-amp output. The *open loop gain* is defined as the ratio of V_o to V_i:

$$\text{OPEN-LOOP GAIN} = \frac{V_o}{V_i}$$

The *closed-loop gain* is defined as the ratio of V_o to V_s:

$$\text{CLOSED-LOOP GAIN} = \frac{V_o}{V_s}$$

The open-loop gain is the gain of the op amp, and this gain is independent of the input and feedback components. The closed-loop gain, however, depends only on the values of the input and feedback components when the closed-loop gain of the circuit is much less than the large open-loop gain of the op amp.

Input Current and Feedback Current

When an input signal (V_s) is applied to the circuit of Fig. 1-2, a current (I_i) flows through the input component, and a voltage (V_i) develops across the input terminals of the op amp. The very high gain op amp amplifies the differential input voltage (V_i) producing an output voltage (V_o) with a polarity opposite to that of V_i. This output is fed back through the feedback component and opposes the input voltage that produced it.

Because the negative feedback signal opposes the input signal, V_i is very small. The higher the gain of the op amp, the smaller is V_i. In fact, for some calculations, V_i can be assumed equal to zero and the inverting input at virtually the same potential as the noninverting input.

The relationship between the input current (I_i) and the feedback current (I_f) is most important. Assuming that V_i is equal to zero*, it follows from Ohm's law that no current can flow into the op amp. By applying Kirchhoff's current law to the inverting-input terminal of the op amp, it can be seen that the input current must be equal and opposite to the feedback current. Extensive use will be made of this important result:

$$I_f = -I_i$$

The Equivalent Circuit

The schematic diagram of the op amp with negative feedback can be simplified by using the preceding results. This simplification is called an *equivalent circuit*. The equivalent circuit can be obtained by recalling that I_i is equal and opposite to I_f, and so Fig. 1-2 can be relabeled as shown in Fig. 1-3.

Since V_i is nearly zero, the inverting input of the op amp can be considered to be at ground potential, and the schematic diagram of Fig. 1-3 is further simplified to the important equivalent circuit of Fig. 1-4.

The equivalent circuit (Fig. 1-4) suggests why op amps with negative feedback are so useful. The input circuit is electrically isolated from the output circuit, yet the current flowing through the input component dictates what current must flow through the feedback component. By choosing different input and feed-

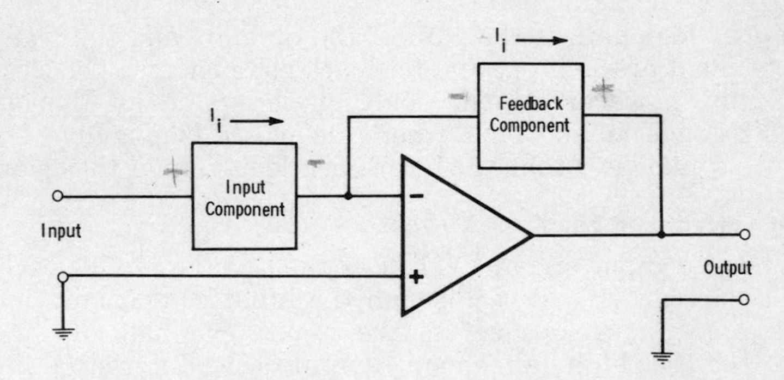

Fig. 1-3. Negative feedback op-amp circuit with equivalent feedback current shown.

---

*If V_i were actually equal to zero, the output of the op amp would also be zero, and our op amp would be about as useful as a blown fuse. In reality, V_i is a very small voltage (usually less than a millivolt). But *for the purpose of calculating input and feedback currents*, very little error is introduced by approximating V_i as equal to zero.

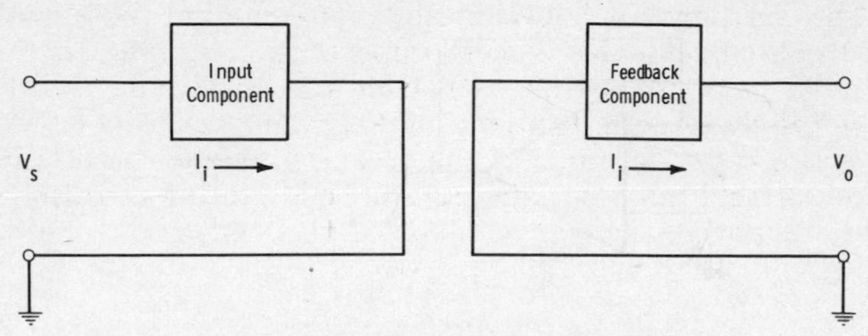

Fig. 1-4. Equivalent circuit of Fig. 1-3.

back components, different circuit functions can be performed. The circuit functions listed in Table 1-1 will be discussed in this chapter, and many more examples will be given in Chapter 5.

Table 1-1. Table of Circuit Functions

Circuit Function	Input Component	Feedback Component
Amplification	Resistor	Resistor
Integration	Resistor	Capacitor
Differentiation	Capacitor	Resistor

NEGATIVE-FEEDBACK CIRCUITS

Most op-amp applications are in negative-feedback circuits. To understand how these circuits work, it is helpful to analyze them in terms of the equivalent circuit of Fig. 1-4. In this section, we will look at four basic negative-feedback circuits using the ideal op amp.

The Inverting Amplifier

An op amp is a very high gain amplifier, but it is rare that so much gain is needed *per se*. This section will show how to use an op amp to build an amplifier with any gain we choose. For an amplifier, both the input component and the feedback component are resistors, as shown in Fig. 1-5A. Fig. 1-5A can be simplified by the equivalent circuit shown in Fig. 1-5B.

The gain (G) of the amplifier circuit is given by:

$$G = \frac{V_o}{V_s}$$

So to determine the gain of the amplifier, we need only to find V_o and V_s, which are given by Ohm's law:

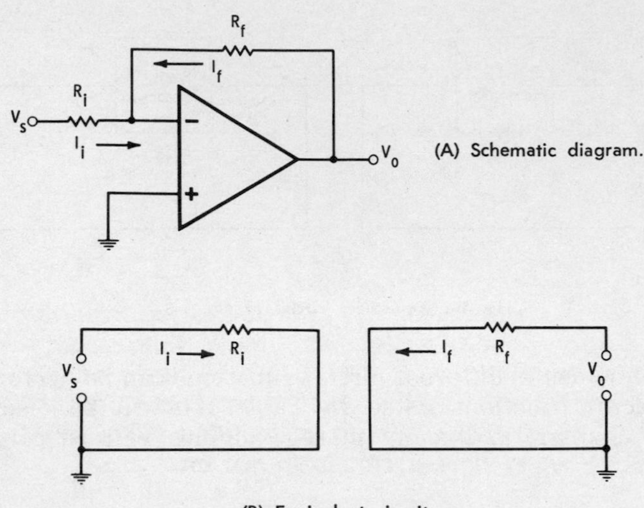

(A) Schematic diagram.

(B) Equivalent circuit.

Fig. 1-5. Inverting amplifier.

$$V_o = I_f\,R_f$$
$$V_s = I_i\,R_i$$

Remembering the important result that:

$$I_f = -I_i$$

The voltage equations become:

$$V_o = -I_i\,R_f$$
$$V_s = I_i\,R_i$$

The gain of the amplifier is then given by:

$$G = \frac{V_o}{V_s} = \frac{-I_i\,R_f}{I_i\,R_i} = -\,\frac{R_f}{R_i}$$

This important result states that the gain of the amplifier is equal to the feedback resistance divided by the input resistance. The minus sign indicates that the output is 180° out of phase with the input, and for this reason the circuit is called an *inverting amplifier*. It can be seen from the equivalent circuit (Fig. 1-5B) that the input impedance of the amplifier is just equal to the input resistance R_i. The output impedance of this circuit is zero since the output voltage is determined by the feedback current which, for the ideal op amp, is not affected by the load.

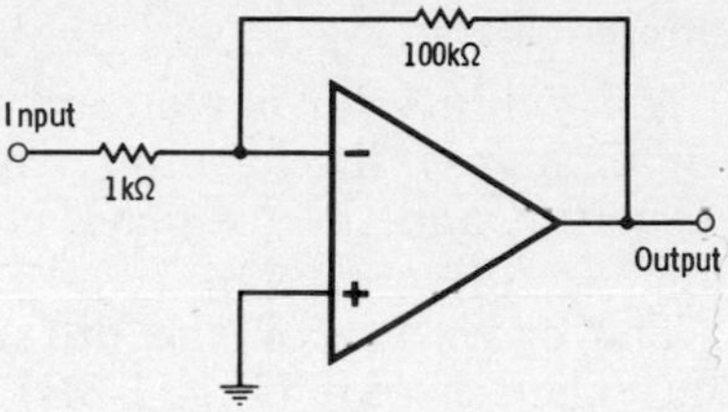

Fig. 1-6. Amplifier with gain of 100.

Now let us see how easy it is to design an amplifier stage using an op amp with feedback. For example, suppose we want to build an amplifier with an input impedance of 1000 ohms and a gain of 100. It takes just three components: (1) an op amp, (2) a 1000-ohm input resistor, and (3) a 100,000-ohm feedback resistor. This circuit is shown in Fig. 1-6.

Integrators

An *integrator* is a circuit that "integrates" (or takes the sum of) the input signal over a period of time. For an integrator, the input component is a resistor and the feedback component is a capacitor, as seen in Fig. 1-7A. The equivalent circuit in Fig. 1-7B shows that the charging current of the capacitor is equal to the current flowing in the input resistor, and that the output voltage is equal to the voltage across the capacitor. The output of the op-amp integrator is proportional to the negative of the integral of the input.

When a constant voltage is applied to the input of the integrator, a constant capacitor-charging current will flow, and the voltage across the capacitor will increase linearly. The capacitor is "integrating" or "adding up" the input voltage. So, for example, when the input to the integrator is a square wave, the output is a triangular wave. The op-amp integrator is most commonly used in analog computers.

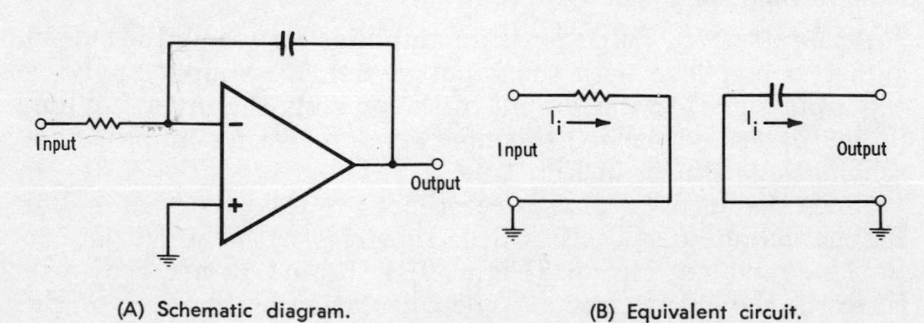

(A) Schematic diagram. (B) Equivalent circuit.

Fig. 1-7. Integrator circuit.

Differentiators

A *differentiator* is a circuit that responds only to differences (or changes) in the input signal. The input component is a capacitor, and the feedback component is a resistor, as shown in Fig. 1-8A.

The operation of the differentiator can be seen from the equivalent circuit (Fig. 1-8B). A capacitor will not pass direct current. Only changes in the input signal will result in current flow through the input capacitor, and so only changes in the input signal will result in current flow through the feedback resistor. The output of the differentiator is equal to the voltage drop across the feedback resistor and is proportional to the negative of the derivative of the input.

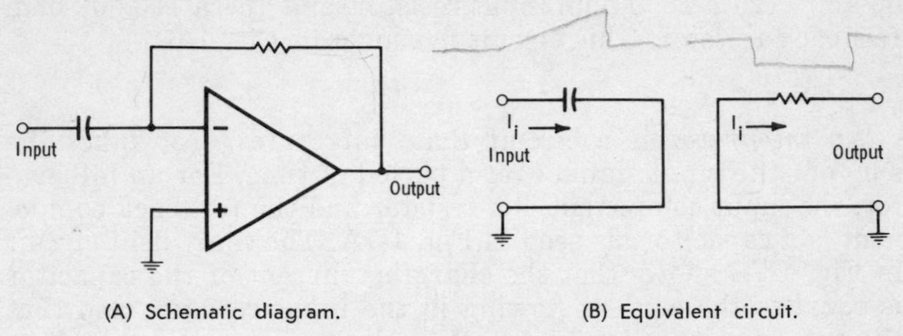

(A) Schematic diagram.

(B) Equivalent circuit.

Fig. 1-8. Differentiator circuit.

A differentiator always does the exact opposite of an integrator. When the input to a differentiator is a triangle wave, the output is a square wave. When the input to a differentiator is a square wave, the output is a series of voltage spikes corresponding to the square-wave voltage changes. The op-amp differentiator is useful in instrumentation for determining the rate of change of voltage variables.

Noninverting Amplifier

A *noninverting amplifier* is an amplifier for which the output signal is in phase with the input signal. By simply applying the input signal to the +input of the op amp, the inverting amplifier described earlier becomes a noninverting amplifier. A schematic is shown in Fig. 1-9A.

Since the output signal is referenced to the grounded end of R_i, the voltage drop across R_i is in series with the output. So in the equivalent circuit (Fig. 1-9B), R_i and R_f are shown in series in the output circuit. The calculation of the gain of the noninverting amplifier is straightforward:

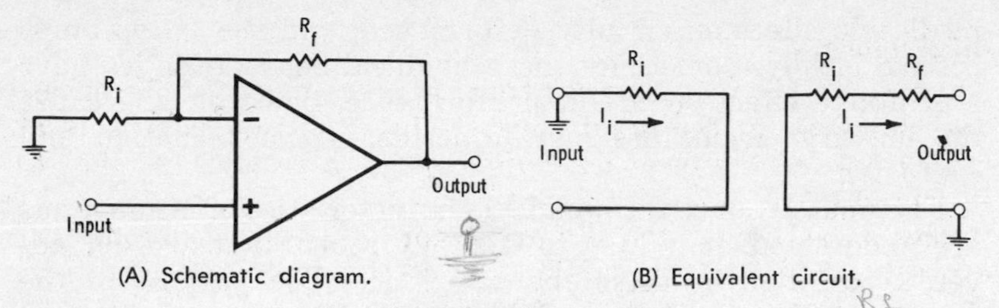

Fig. 1-9. Noninverting amplifier.

$$G = \frac{V_o}{V_s} = \frac{I_i\,(R_i + R_f)}{I_i\,R_i} = \frac{R_i + R_f}{R_i}$$

The gain of the noninverting amplifier is equal to the sum of the input and feedback resistances, divided by the input resistance.

Summary

Four circuits have been described to illustrate the use of ideal op amps with negative feedback. A summary of the output waveforms of these circuits for a square-wave input is given in Fig. 1-10. In these four examples it can be seen that the analysis of op-amp circuits with negative feedback can be greatly simplified by using the equivalent circuit of Fig. 1-4.

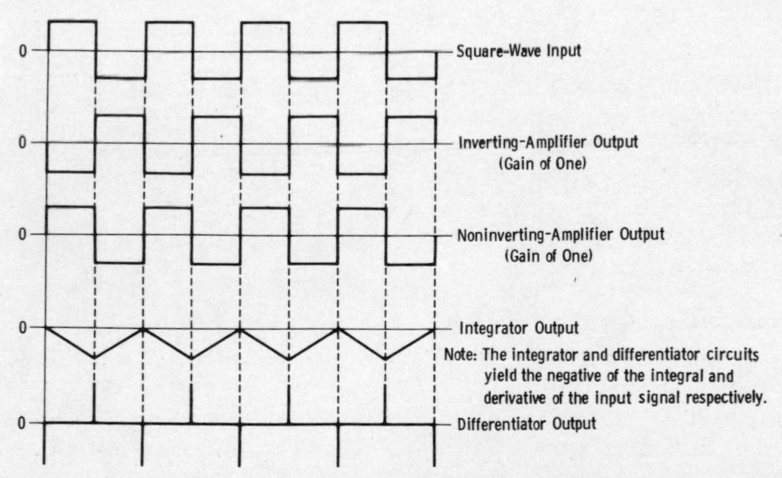

Fig. 1-10. Summary of output waveforms for a square-wave input.

NONIDEAL CHARACTERISTICS

IC op amps, like all real op amps, do have some nonideal characteristics. Often these nonideal characteristics have a negligible effect on circuit performance, and the IC op amps behave ideally. Sometimes these nonideal characteristics have a profound effect on circuit performance, however, and cause the unwary circuit designer unnecessary consternation and confusion.

Fortunately, once the nonideal characteristics of IC op amps are understood, it is easy to design circuits that are not adversely affected by these characteristics. The purpose of the next three chapters is to explain how an IC op amp works and just exactly what these nonideal characteristics are.

CHAPTER 2

IC Electronics

An elementary knowledge of IC electronics is essential to a thorough understanding of IC op amps. This chapter is an introduction to the fundamentals of IC electronics—semiconductor physics, semiconductor devices, and integrated-circuit technology.

SEMICONDUCTOR PHYSICS

A *semiconductor* is a crystalline material with specific electrical properties. Silicon and gallium arsenide are the most popular semiconductors in use today. All integrated circuits presently being built use silicon.

Fig. 2-1 shows a silicon ingot that has been grown for use in integrated circuits. These ingots are usually several inches in diameter and several feet long. The ingots are sliced into wafers about 0.01 inch thick. These silicon wafers are used as the starting material in the IC fabrication process.

Semiconductor Conduction

Metals provide a very simple illustration of the process of electrical conduction. When a voltage is applied across a metal, a "sea of electrons" drifts in response. Even though high currents may flow in a metal, the average velocity of the electrons is slow (typically less than ½ inch per second) because of the large number of electrons participating in the conduction. While the "sea-of-electrons" idea provides a clear picture of conduction in metals, it needs modification before it can be

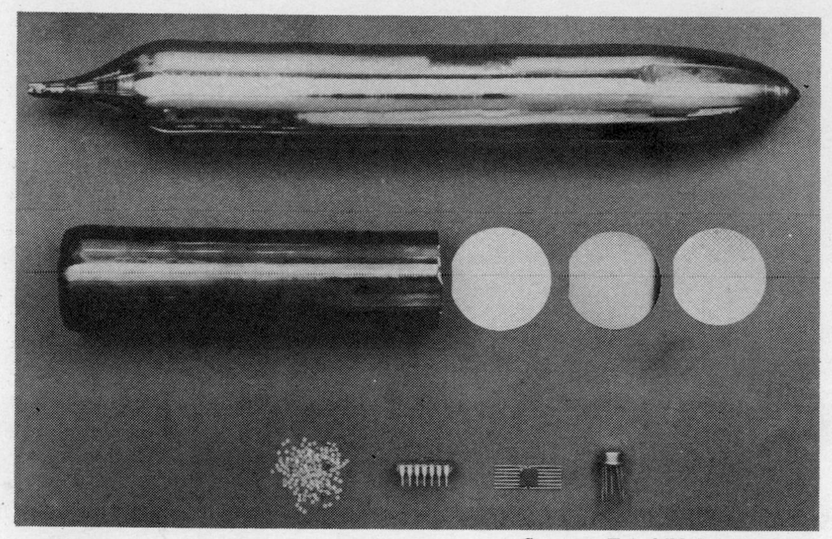

Courtesy Fairchild Semiconductor

Fig. 2-1. The monolithic fabrication process from silicon ingot to packaged IC.

extended to semiconductors. Just as in metals, though, electrons are the only type of charge that contributes to semiconductor conduction.

Electrons in semiconductors have certain specific energies. The allowed energies are grouped, and these groups of energy levels are called *energy bands*, or just *bands*. In nature, systems tend to exist in the lowest energy state. Consequently, electrons tend to occupy mostly the lower energy bands. The two highest energy bands containing electrons are called the *valence band* and the *conduction band* (Fig. 2-2), and it is these bands that are important to conduction in semiconductors.

The conduction band is the higher of these two bands. It is only slightly filled. The electrons in this energy band act quite similarly to electrons in a metal during conduction. When a voltage is applied across a semiconductor, these electrons drift in response, similar to the sea of electrons. The current identified with the flow of electrons in this band is called *electron current*, and the *carriers*, or type of charge said to be "carrying" the current, are electrons.

The valence band is just below the conduction band. The energy levels in this band are mostly filled. There are very few vacant energy levels available. It has been shown with quantum mechanics that these vacancies (or *holes*) behave very much like the electrons in the conduction band, except that they act as though they are positively charged. This can

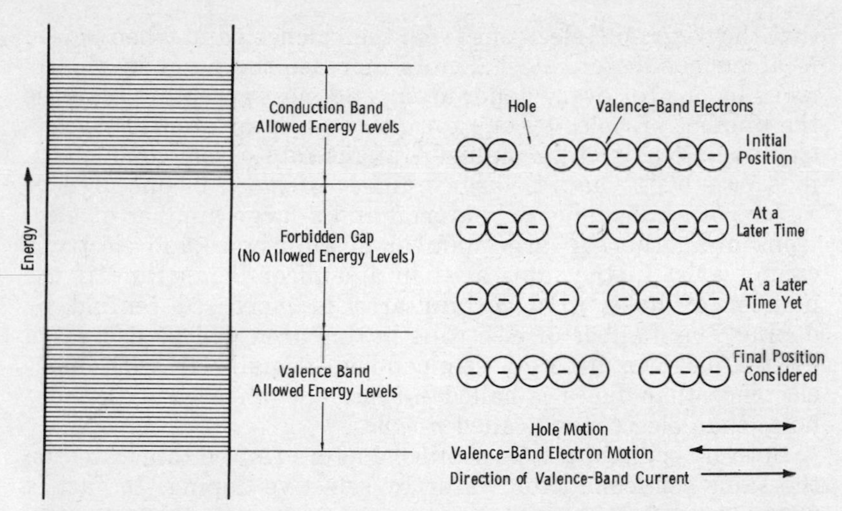

Fig. 2-2. Semiconductor energy levels.

Fig. 2-3. Conceptual diagram of valence-band conduction.

be seen more clearly by looking at conduction in the valence band, called *hole conduction*. When a voltage is applied across a semiconductor, electrons in the valence band jump into nearby vacancies, leaving a vacancy at the energy level from which they came. As seen in Fig. 2-3, the electrons are moving toward the left, but from another point of view, the vacancy, or hole, has moved toward the right. Thus, conduction in the valence band can be considered in terms of holes moving in the direction opposite to the electrons in the band.

In summary, when a voltage is placed across a semiconductor, there are two forms of current: (1) current due to electrons in the conduction band, and (2) current due to holes in the valence band. Unless otherwise noted, the term *electrons* will be used to describe only the current carriers in the conduction band, and the conduction-band current will be referred to as *electron current*. Similarly, the term *holes* will be used to describe current carriers in the valence band, and the valence-band current will be referred to as *hole current*.

N-Type and P-Type Semiconductors

The concentration of electrons and holes in a semiconductor can be controlled. Two classes of impurity atoms are used to control the number of electrons and holes in semiconductors: (1) *donors* and (2) *acceptors*. Donor atoms have excess electrons which they are willing to "donate" when placed in a semiconductor. Acceptor atoms are the opposite of donors in that they "accept" electrons from the valence band when placed in a semiconductor. Most donors increase the number of electrons by one for every donor atom, and most acceptors increase the number of holes by one for every acceptor atom. By *selective doping* (injecting impurity atoms into a semiconductor), it is possible to have a large number of holes in one area of a piece of semiconductor material and a large number of electrons in another. A large number of both holes and electrons cannot exist in the same area in a semiconductor, and if the number of holes in a certain area is increased tenfold by doping, the number of electrons in this area will be decreased by tenfold. An area of semiconductor material with more electrons than holes is called *n-type,* and an area with more holes than electrons is called *p-type.*

Both n-type and p-type semiconductors can be fabricated on the same semiconductor wafer by selective doping. In fact, a p-type semiconductor can be changed to an n-type by *counterdoping* acceptor-doped material with a greater number of donor atoms. Counterdoping can likewise be used to change n-type material to p-type material.

Semiconductor Doping

Semiconductors are typically doped at two different times during IC fabrication. During the growth of the semiconductor crystals, a *dopant* is added to make the entire crystal n-type or p-type. After the crystal is sliced into wafers, it may be doped by a process called *diffusion.*

Diffusion is simply the migration of atoms from an area of high concentration to an area of low concentration. Many people have observed atomic diffusion when smelling perfume across the room from its source. The diffusion-doping process is carried out by passing a dopant-rich gas over the surface of a semiconductor wafer that has been heated to about 2000 °F. The dopant atoms from the gas are absorbed on the surface of the semiconductor and diffuse deep into the semiconductor material.

SEMICONDUCTOR DEVICES

The Diode

The *diode* is the simplest integrated-circuit structure. It consists of an n-type doped-semiconductor area adjacent to a p-type doped-semiconductor area, as shown in Fig. 2-4A. The diode *V-I* (voltage-current) characteristic curve and schematic symbol are also shown in Fig. 2-4. A diode may be

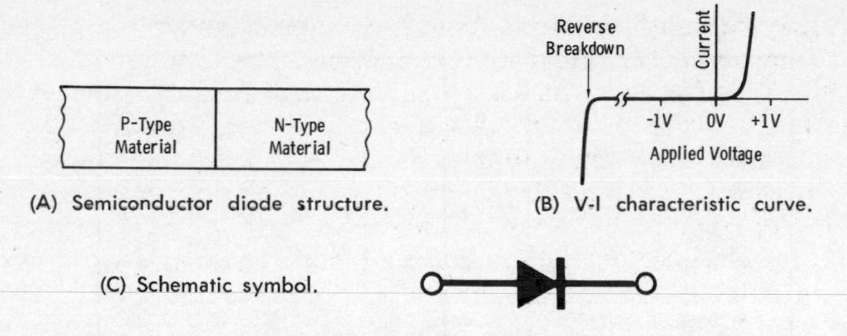

(A) Semiconductor diode structure.

(B) V-I characteristic curve.

(C) Schematic symbol.

Fig. 2-4. Semiconductor diode data.

fabricated by counterdoping part of an n-type silicon wafer to be p-type through a diffusion process.

A diode ideally passes current in only one direction, as shown diagrammatically in Fig. 2-5. In order to understand how this diode action occurs in a semiconductor, it is first necessary to understand a balance that goes on in semiconductors.

This balance exists between the two types of current that can exist in a semiconductor: (1) *drift current* and (2) *diffusion current*. Drift current is caused by the flow of carriers in an electric field. Diffusion current is caused by the migration (or diffusion) of carriers from an area of high carrier concentration to an area of lower concentration of the same type of carrier. Carrier diffusion is similar to the atomic diffusion of impurity atoms discussed previously; however, there is a major difference between these two types of diffusion: Carriers at room temperature diffuse readily, whereas dopant atoms at room temperature diffuse at a rate which, for all practical purposes, is zero. This is fortunate, since it is undesirable to have the dopant atoms move once they are initially placed.

When there is a difference in hole concentration between two adjacent areas in a semiconductor, a diffusion current will flow from the area of high hole concentration to the area of low hole concentration. Since the holes are diffusing from an electrically neutral area, an electric field is created between the holes that have diffused and the negatively charged area they

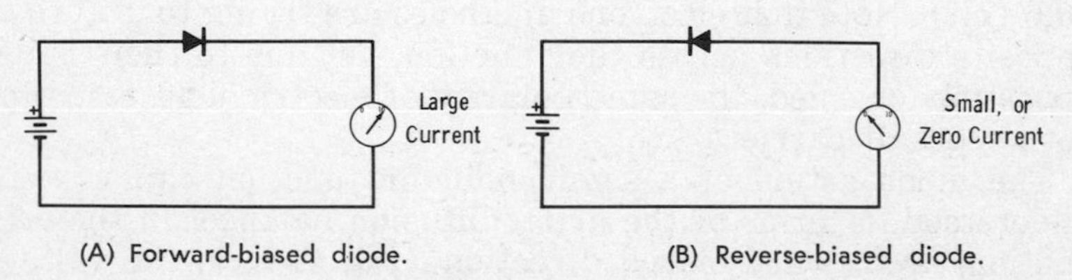

(A) Forward-biased diode.

(B) Reverse-biased diode.

Fig. 2-5. Diagrammatic representation of diode action.

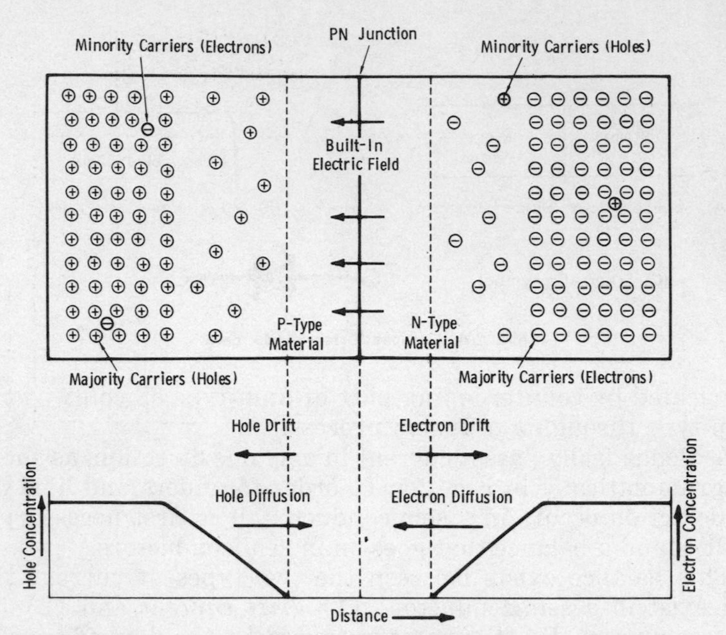

Fig. 2-6. Pictorial drawing of semiconductor diode, and plot of majority-carrier concentration.

created in leaving. This electric field causes a drift current to flow, as shown in Fig. 2-6. At equilibrium, the diffusion current is equal and opposite to the drift current, and there is no net current flow.

The balance between drift and diffusion current is very important at the junction between the p-type and the n-type material in a semiconductor diode. On the p-type side of the junction there are a large number of holes, while on the n-type side there are few. Thus, there is a balance between drift and diffusion currents for both the electrons and the holes. The amount of diffusion current that would flow, if it were not restrained by the electric field, would be several thousand amperes for a typical diode. Yet in a silicon diode, this current is restrained by the electric field, which is only 600 millivolts (0.6 volt). Note that electrons and holes are trying to diffuse in opposite directions across the junction, yet due to their being oppositely charged, the same polarity of electric field restrains both types of carriers.

The diode action of a semiconductor junction can now be understood in terms of the drift/diffusion balance. In the conduction, or forward-biased direction (Fig. 2-7A), the voltage applied from a circuit to the diode decreases the built-in electric

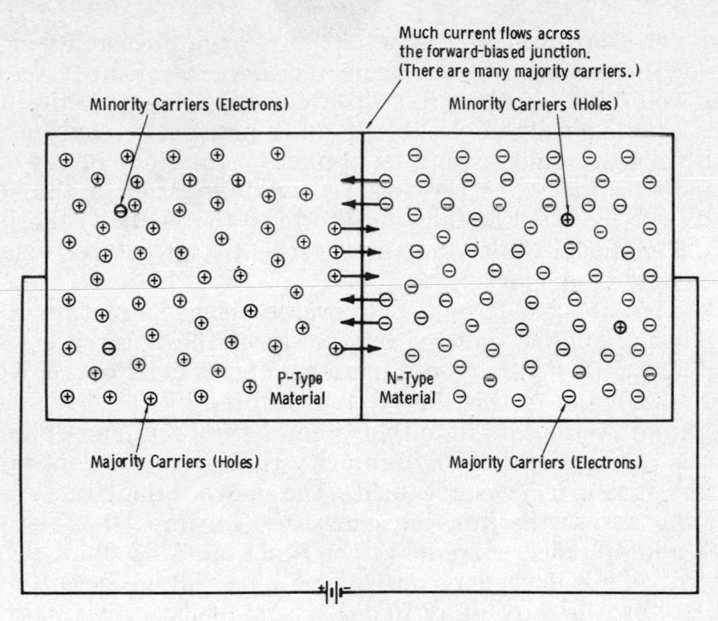

(A) Forward-biased diode.

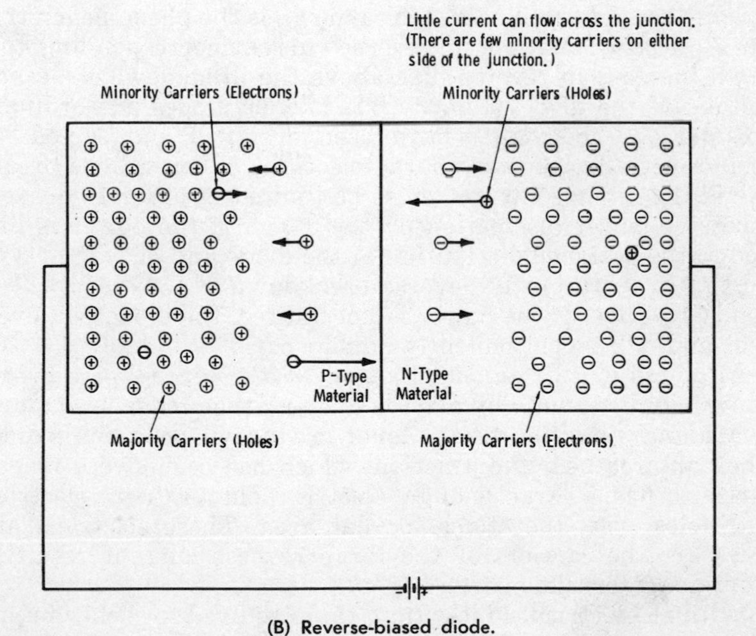

(B) Reverse-biased diode.

Fig. 2-7. Pictorial representation of diode action.

field, releasing part of the potentially large diffusion current. For a silicon diode, several thousand amperes of diffusion current would flow if 600 millivolts were applied across the junction. This is nearly impossible to do in practice, because a very high voltage would have to be applied to the leads of the diode in order to overcome the resistive voltage drop in the semiconductor material as the current flows to and from the junction. The diode would probably overheat long before this current could be achieved.

In the nonconducting, or reverse-biased direction (Fig. 2-7B), the applied voltage acts to upset the balance between drift and diffusion currents in favor of the drift current. The current flowing in this direction is limited by the diffusion of the minority carriers on either side of the junction. Upon diffusing to the junction, the minority carriers are swept across by the bias-aided electric field. These few minority carriers drifting across the junction constitute an imbalance between drift and diffusion currents at the junction. This small reverse current is not dependent on reverse-bias voltage because it is limited by the rate of diffusion of minority carriers to the junction.

If the reverse-bias voltage is large, the diode will undergo *reverse breakdown*. Reverse breakdown is the phenomenon that causes a large increase in reverse current corresponding to a small increase in reverse bias above the breakdown, or *zener*, voltage of the diode (Fig. 2-4B). (Zener diodes are ordinary pn-junction diodes that have been designed and sorted for their reverse-breakdown characteristics.) When a diode breaks down, the reverse current must be limited in order to prevent excessive power dissipation by the diode. If the power is limited to the designed dissipation of the diode and package (typically $\frac{1}{4}$ to $\frac{1}{2}$ watt), the reverse breakdown of the diode will be nondestructive. If the power is not limited, the diode will overheat and will be permanently damaged.

Although it is often called *zener breakdown*, reverse breakdown in a diode actually may be due to either zener breakdown or *avalanche multiplication*. Zener breakdown occurs in a diode when an area near the junction, which has been swept free of carriers, has a large enough electric field to "rip" electrons and holes from the atoms in that area. These electrons and holes are the carriers of the large reverse current resulting from zener breakdown.

With an electric field less than that required for zener breakdown, avalanche multiplication can occur. Avalanche multiplication is the effect that causes an increase in the reverse current by the collision of the carriers constituting the reverse current (which are accelerated by the large electric field near the junction) with the crystal lattice near the junction. These collisions "knock loose" hole-electron pairs from the silicon atoms which, depending on the location of the original collision, may be accelerated to cause more carrier-generating collisions. It is the electrons and holes knocked off by collisions that constitute carriers of the large reverse current due to avalanche breakdown.

It is interesting to note that the variation of the breakdown voltage with temperature is in a different direction for the two effects. Since zener breakdown usually occurs with silicon diodes that break down at less than 6 volts, while avalanche breakdown occurs with silicon diodes that break down at 6 volts or more, a diode with approximately a 6-volt breakdown can be used as a stable voltage reference that has little variation with temperature.

To summarize, silicon diodes conduct large currents in the forward-biased direction with roughly 600 millivolts applied to them. In the reverse-biased direction, only a little current flows, unless the bias is large enough to cause reverse breakdown. Reverse breakdown will not harm a diode as long as the power dissipated in the diode is limited to the designed level.

The Bipolar Transistor

The bipolar transistor is a semiconductor device that can be used to proportionally control large currents with small currents. There are two basic types of bipolar transistors: (1) the npn type and (2) the pnp type. Fig. 2-8 shows the schematic symbol and semiconductor structure for each type of bipolar transistor.

The operation of a bipolar transistor with a current gain (β) of 100 is shown in Fig. 2-9 for both types of transistors. It can be seen from Fig. 2-9 that the current allowed to flow into the collector is 100 times that put into the base.

From the structure of the bipolar transistor, it becomes apparent that the transistor actually is two closely spaced diodes arranged "back-to-back." The transistor can be explained in terms of a coupling between the forward-biased, emitter-base diode, and the reverse-biased, base-collector diode. A little further understanding of how diodes work beyond that given in the previous section is necessary before *transistor action* (the coupling between the diodes connected "back-to-back") can be fully understood.

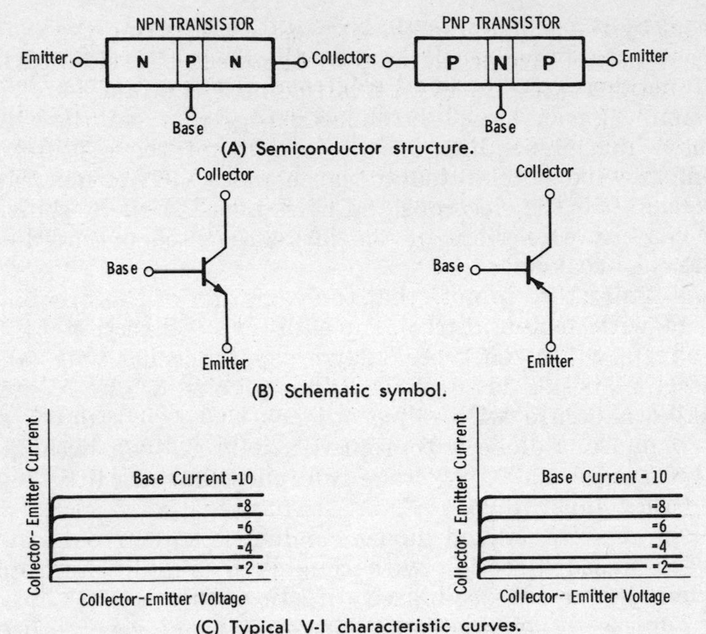

(A) Semiconductor structure.

(B) Schematic symbol.

(C) Typical V-I characteristic curves.

Fig. 2-8. Bipolar transistor data.

In a forward-biased junction, the majority carriers on both sides of the junction diffuse across the junction. Once across the junction, these carriers can be considered minority carriers. Thus, holes diffuse across the junction from the p-type side to the n-type side to join the holes on the n-type side. For small forward currents, called the *low-level injection* case, the hole concentration on the n-type side of the forward-biased junction will remain much less than the electron concentration. This increase in the number of minority carriers on both sides of the junction disturbs the proportion of electrons and holes set by the doping. It is because of this imbalance that electrons on the n-type side tend to leave the conduction band and

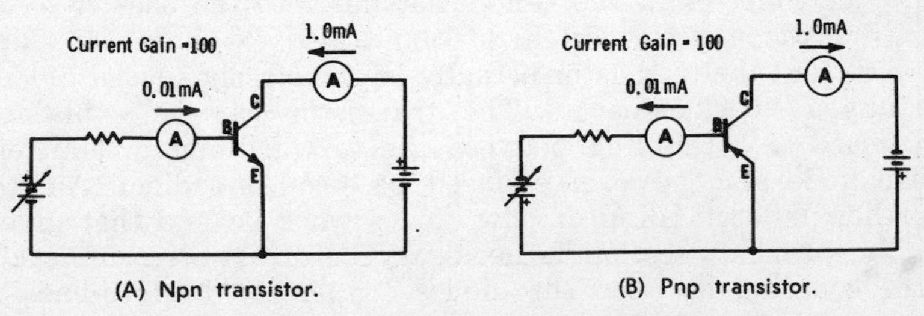

(A) Npn transistor.

(B) Pnp transistor.

Fig. 2-9. Simplified illustration of bipolar transistor action.

fill up the excess holes as the holes diffuse away from the junction they just crossed. It is generally true that whenever the proportioning of holes and electrons is altered from that set by the doping, the semiconductor will act in the direction to restore the proper proportioning. The process of an electron leaving the conduction band to go to the valence band, or vice versa, is called *recombination.* Excess electrons on the p-type side of this forward-biased junction also recombine. The length of time it takes before these excess minority carriers recombine is called the *minority carrier lifetime.*

By heavily doping the n-type side of the junction, and lightly doping the p-type side, it is possible to have the electron diffusion current across the junction be much greater than the hole diffusion current across the junction. The operation of this n^+p junction ("+" refers to high doping density, not charge) could be considered as only having electrons *injected* into the p-type side, neglecting the small hole current diffusing across the junction.

The operation of an npn transistor can be easily understood if the n^+pn transistor shown in Fig. 2-10 is considered. The forward-biased n^+p diode injects electrons into the p-type side,

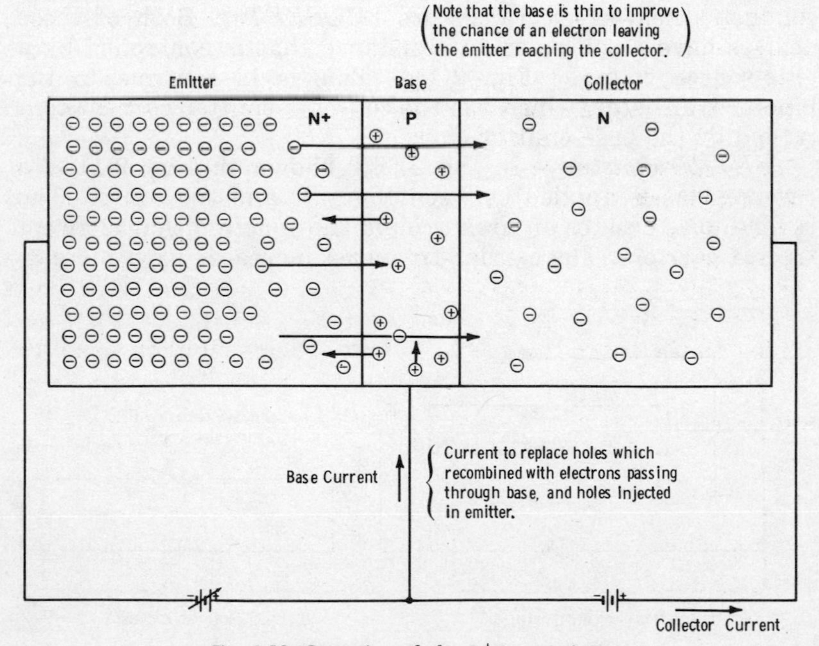

Fig. 2-10. Operation of the n^+pn transistor.

called the base of the npn transistor. If the lifetime of the electrons is long enough (i.e., they do not have time to recombine), the electrons diffuse across the base to the other junction, which is reverse biased. These electrons that have diffused across the base are swept across the reverse-biased, base-collector junction into the collector, where they are once again majority carriers.

There are two components of the base-emitter current that must be supplied in order to allow the collector-emitter current to flow: (1) the small hole current which diffuses across the forward-biased base-emitter junction, and (2) the current which recombines with the electrons that did not have sufficient lifetime to diffuse all the way across the base. Both of these currents are small. For a high-β transistor, the base-emitter current can typically be 1/500 that of the collector-emitter current.

Other Types of Transistors

The bipolar transistor is not the only type of transistor. *Unipolar*, or *field-effect* transistors (*FETs*) constitute the remainder of devices that are currently called transistors.

There are two basic types of field-effect transistors: (1) junction field-effect transistors (*JFETs*) and (2) metal-oxide semiconductor field-effect transistors (*MOSFETs*). Both of these devices have a source-drain resistance that is controlled by a gate-source voltage (Fig. 2-11). This is in contrast to the bipolar transistor, which has the collector-emitter current controlled by the base-emitter current.

JFET Transistors—In the JFET shown in Fig. 2-11A, a reverse bias is applied between the gate and the source. The reverse bias creates an area around the junction that is swept free of carriers. The carrier-free area increases with increas-

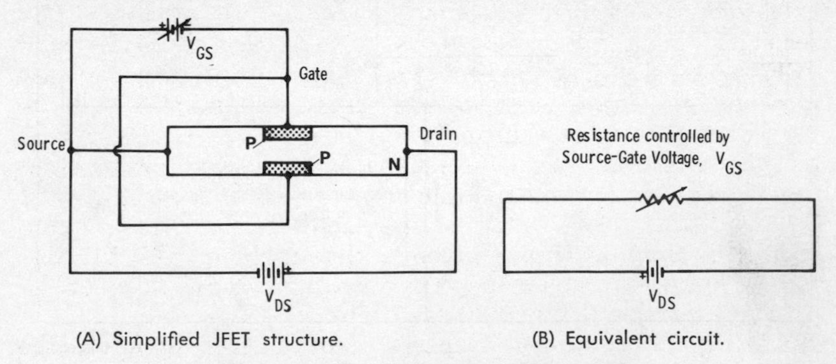

(A) Simplified JFET structure. (B) Equivalent circuit.

Fig. 2-11. Simplified illustration of JFET operation.

Fig. 2-12. Typical JFET characteristic curves.

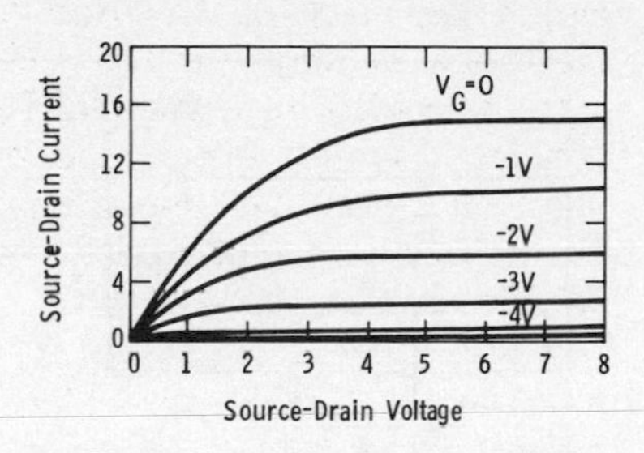

ing reverse bias. Only the section of the channel that has carriers participates in conduction. Thus, increasing the reverse bias increases the source-drain resistance by narrowing the effective channel width. Applying a voltage across the source-drain resistance further reverse biases the gate-channel junction, narrowing even more the portion of the channel with carriers. If a sufficiently large source-drain voltage is applied, the increasing source-drain resistance prevents a further increase in the source-drain current as the source-drain voltage is increased. The V-I characteristic curves for the JFET (Fig. 2-12) are similar to those for a pentode vacuum tube.

MOSFET Transistors—The MOSFET is quite similar to the JFET. Biasing the aluminum gate of the MOSFET negatively (Fig. 2-13) will attract holes to the surface of the silicon and repel electrons away from the surface. By heavily biasing the gate, more holes than electrons will be at the surface, effectively changing the doping from n-type to p-type. This changed (or *inverted*) area provides a resistive conduction path between the source and drain. The depth of the channel, hence the source-drain resistance, is dependent on the gate-source bias.

BIFET Transistors—The BIFET transistor, which is used as the input amplifier stage in the popular 080, 081, and 082 series of operational amplifiers, is a specially constructed JFET. The JFET transistor has many properties that are ideal for the input stage of an operational amplifier. Since very little current is required at the input of the device to control the source-to-drain current, the JFET has very high input impedance. Unfortunately, the simple JFET shown in Fig. 2-11 is difficult to fabricate at the same time as bipolar transistors. A special name is given to the bipolar-JFET, fabrication-compatible transistors. They are called *BIFETs*, or bipolar-compatible, field-effect transistors.

BIMOS Transistors—A special name is also given to the MOS transistors that are compatible with bipolar transistor

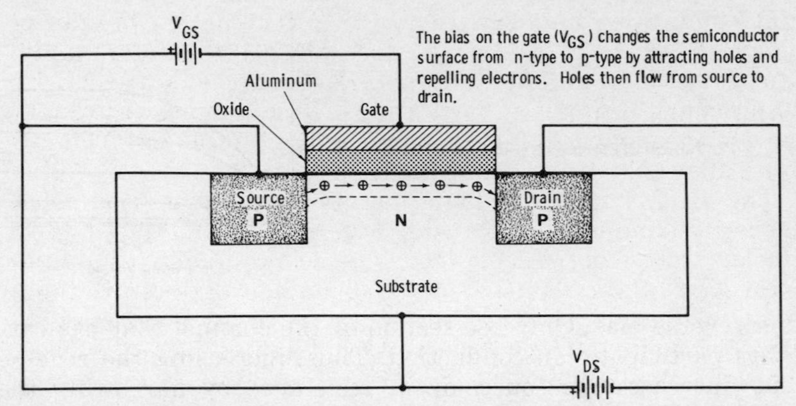

Fig. 2-13. Simplified illustration of MOSFET operation.

fabrication. These devices are called *BIMOS* transistors, or bipolar-compatible, metal-oxide-semiconductor transistors.

CMOS Transistors—There are two types of MOS transistors: (1) n-channel and (2) p-channel. When fabricated together in a complementary-symmetry arrangement, they are called *CMOS* devices. A thorough discussion of these devices is presented in *Understanding CMOS Integrated Circuits,* by Melen and Garland, published by Howard W. Sams & Co., Inc., 1975.

Transistor Comparisons

Given the prospect of using op amps with either a MOSFET, JFET, or bipolar input amplifier stage, the designer can get a feel for the performance differences of the devices by considering their basic structures.

The bipolar device requires current as a basic part of its mechanism of operation. This input current is to provide for the current injected into the emitter and for the recombination current in the base of the transistor. It would not be surprising then to learn that this transistor forms an undesirable input stage when the circuit requires that the device input currents be small. The bipolar device does have low thermal noise and high transconductance. Therefore, it might be advisable to use these devices in low-noise applications or in low-power, high-gain applications in integrated circuits that effectively utilize the high transconductance of the bipolar transistor.

The JFET (and hence the BIFET) device has only a secondary requirement for current at the input gate. This current is the reverse bias current of the pn-junction diode that comprises the input structure of the device. This reverse current doubles in value for every 10° C increase in the device temperature. As a result, the current can increase to significant values at high temperatures. For example, a JFET with 20 picoamperes input current at 30° C room temperature may have 82 nanoamperes input current at the typical 150° C maximum allowable device temperature.

The BIFET amplifier input stage does have good noise properties and, in some cases, superior noise properties to bipolar transistor input stages. These lower noise properties arise when the circuit can take advantage of the low input noise current of the BIFET compared with the bipolar amplifier. This noise current is small because the absolute value of the total input current to the BIFET amplifier is small compared with the bipolar amplifier.

The BIMOS, CMOS, and MOS transistor amplifier stages are characterized by even lower values of input bias and noise currents than the BIFET. In fact, the input dc current would theoretically be zero if it were not for the bias current associated with the zener protection diodes usually connected to the input gate of the devices. This input protection zener is typically included to prevent damage to the oxide insulation by static electricity. Currents resulting from static electricity can be inadvertently injected into the MOS gate during typical circuit assembly if precautions are not taken. These currents can easily rupture the insulating oxide if protective zeners are not used.

The BIMOS and CMOS input stages are desirable when the absolute minimum value of input currents is desired. However, since the currents in the device flow predominantly at the semiconductor/oxide interface, the BIMOS and CMOS amplifiers tend to have more noise voltage referred to the input resulting from the injection and trapping of the electrons in the oxide layer. This injection of a charge into (and release of charge from) the oxide results in a noise that has a large low-frequency component. This low-frequency noise is popularly called *1/f noise* because when the frequency spectrum is plotted, this component of the noise decreases by a factor of approximately ten for every factor of ten increase in frequency.

THE MONOLITHIC FABRICATION PROCESS

This section explains how integrated circuits containing diodes and transistors are built. The IC fabrication process is called *monolithic* because the integrated circuit is fabricated on a single piece of silicon.

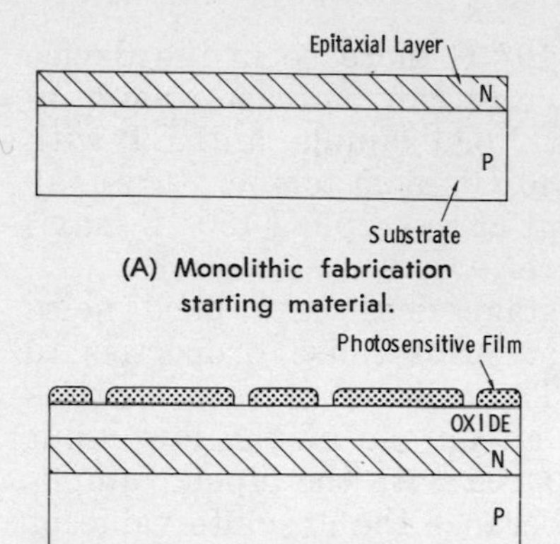

(A) Monolithic fabrication starting material.

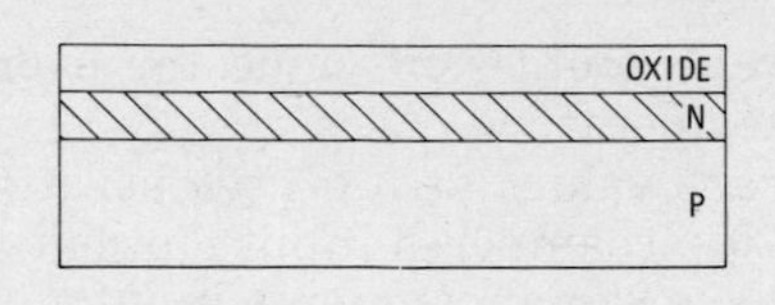

(B) First step is oxidation of the silicon wafer.

(C) A photosensitive film is applied, exposed, and developed on the oxide surface (using a mask to determine the pattern).

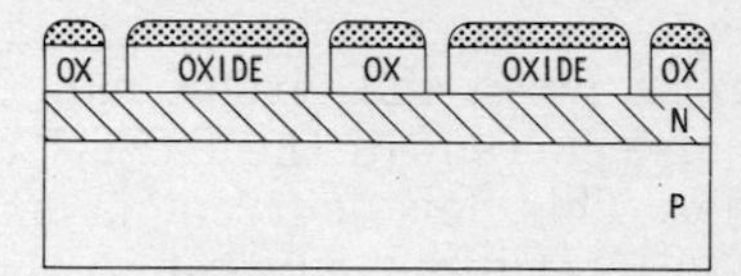

(D) Oxide is etched through windows in the photosensitive film.

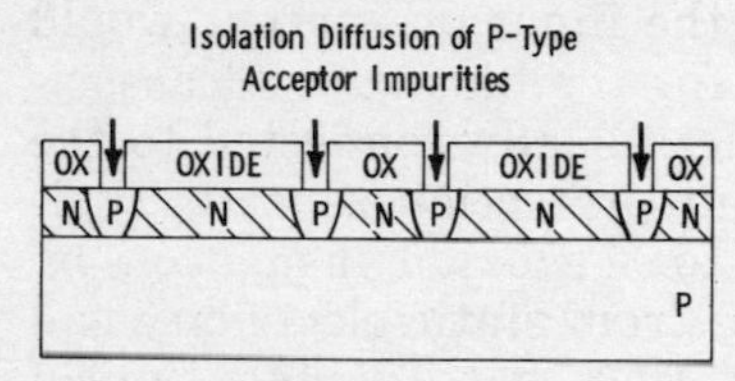

(E) Impurity atoms are diffused into the epitaxial layer through windows in the oxide.

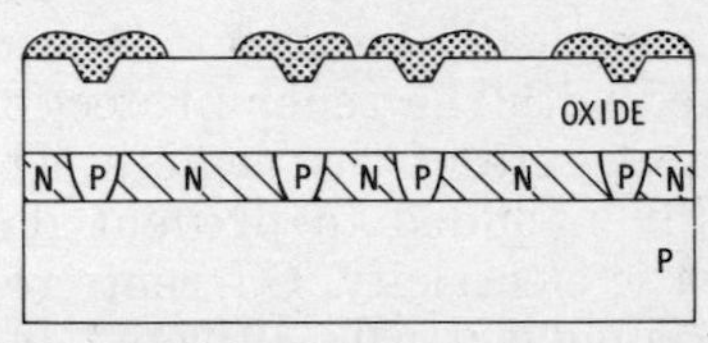

(F) Additional oxide is grown, a photosensitive film is applied, exposed, and developed (using a mask to determine the pattern).

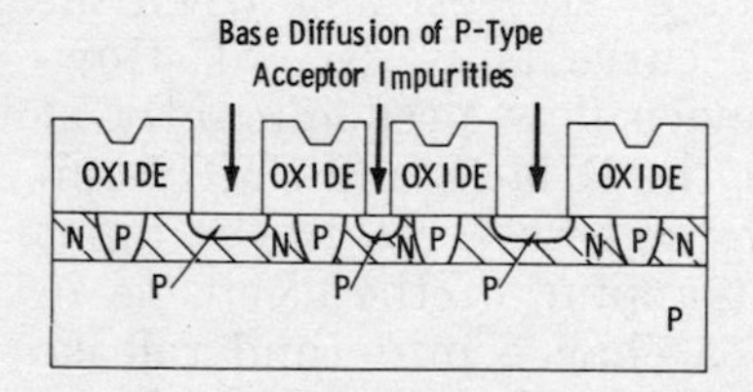

(G) Oxide is etched through windows in the film, and p-type impurity atoms are diffused into the epitaxial layer through the windows in the oxide.

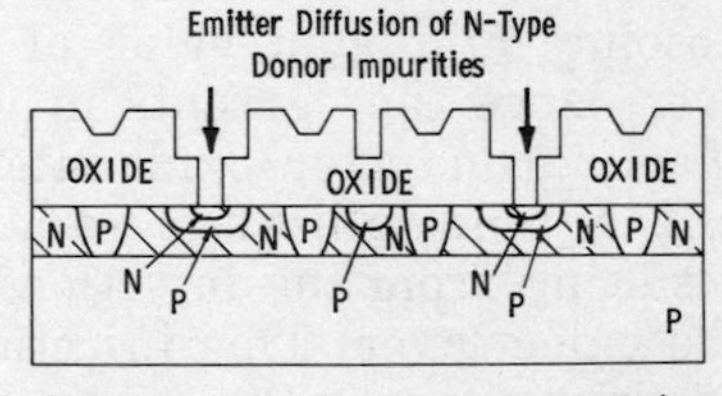

(H) Additional oxide is grown, windows are etched in the oxide using the photosensitive film, and impurity atoms are diffused into the epitaxial layer through the windows in the oxide.

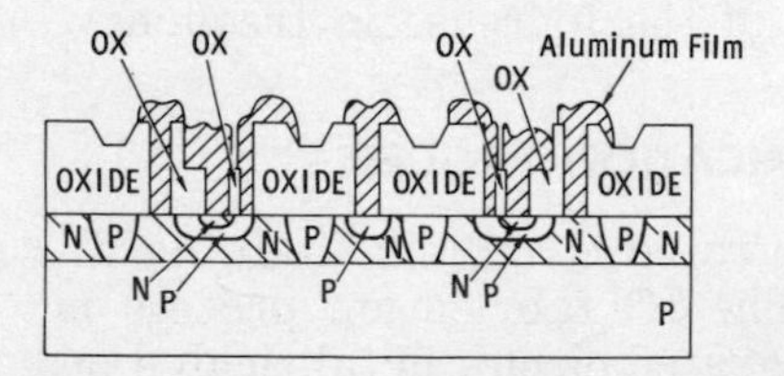

(I) An aluminum film is evaporated everywhere on the surface and selectively removed to create contacts to the semiconductor areas and to the interconnecting wires. (A photosensitive film and an etchant are used to achieve this.)

Fig. 2-14. The monolithic fabrication process.

The Process

A typical monolithic fabrication starts with a p-type silicon wafer with a thin n-type layer on top, as shown in Fig. 2-14A. The p-type bulk is called the *substrate*, and the thin n-type layer is called an *epitaxial layer*. The doping atoms used in both these areas are chosen for their slow atomic-diffusion rates, even at high temperatures, so that the dopant atoms can be assumed not to move with subsequent processing.

The first processing step (Fig. 2-14B) is the oxidation of the silicon wafer by heating it in a high-temperature furnace and simultaneously passing oxygen across the surface of the semiconductor. The oxidized wafer is then covered with a photographic film (which is applied in liquid form) after the wafer has cooled to room temperature. A photographic negative plate, called a *mask* (Fig. 2-15), is placed on top of the silicon wafer, and the film is exposed and developed. The areas of the film exposed to light by the mask remain, while the other areas are removed, as seen in Fig. 2-14C. The wafer is next immersed in an acid which removes only the silicon oxide, and only from areas not covered by the film. Once the remainder of the film is removed, the silicon wafer is placed in a p-type diffusion furnace which dopes the silicon through the windows just etched (as well as doping the top of the silicon oxide). Note that the mask has been used to determine the areas of the semiconductor to be doped. This first diffusion is called the *isolation diffusion* because it separates the n-type epitaxial layer into *isolation islands*. These islands constitute the separation of the components used in an integrated circuit. Transistors, resistors, and capacitors are built in their own isolation islands.

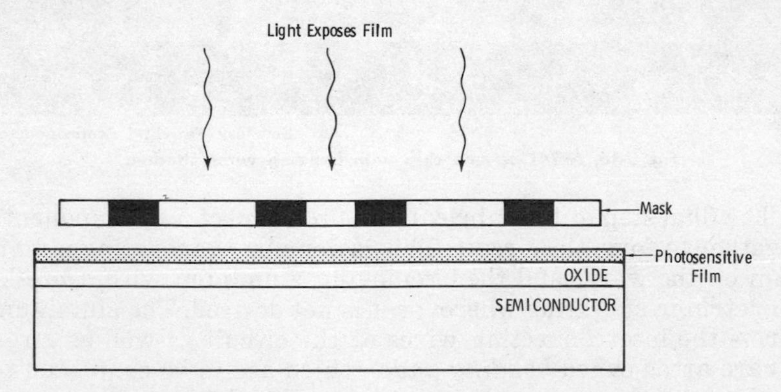

Fig. 2-15. Masking procedure for integrated-circuit fabrication.

The base diffusion (Fig. 2-14G) and the emitter diffusion (Fig. 2-14H) are made in an oxidation-masking-diffusion process similar to the isolation diffusion. The length of time the diffusion is carried on is varied in each case to vary the diffusion depths. Note that each successive diffusion will be more heavily doped in order to effectively counterdope the previous diffusion. The emitter diffusion, being the final diffusion, is the most heavily doped, as it should be to get the n^+p junction discussed in the section on the bipolar transistor.

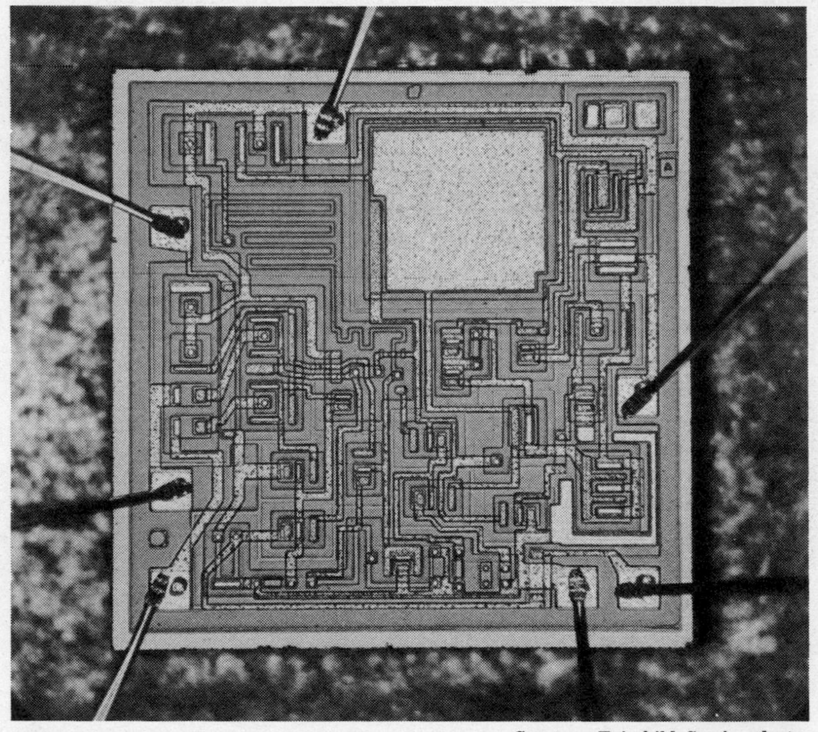

Courtesy Fairchild Semiconductor

Fig. 2-16. A 741 op-amp chip with bonding wires attached.

The final step of the fabrication is to connect the components together to form the circuit. This is done by evaporating aluminum on the wafer and then removing aluminum with a masking/etching step from wherever it is not desired. The aluminum forms the interconnecting wires of the circuit as well as large square areas called *bonding pads*, which are to be connected to the leads of the package it is put into (Fig. 2-16).

Three diffusions and five masking operations have been used to create an integrated circuit. Typically, one wafer has several hundred integrated circuits fabricated on it simultaneously. The masks determine the circuit that is built. The number of circuits built on a wafer depends on the number of times the pattern is repeated on the mask, the area of each pattern, and the area of the silicon wafer.

An automatic testing machine probes each circuit individually with small electrodes, and performs many tests before the wafer is broken up into individual circuits. Usually, fewer than one third of the circuits are good at this point. The good ones, which have been marked by the testing machine, are put into packages such as those shown in Figs. 2-1 and 2-17.

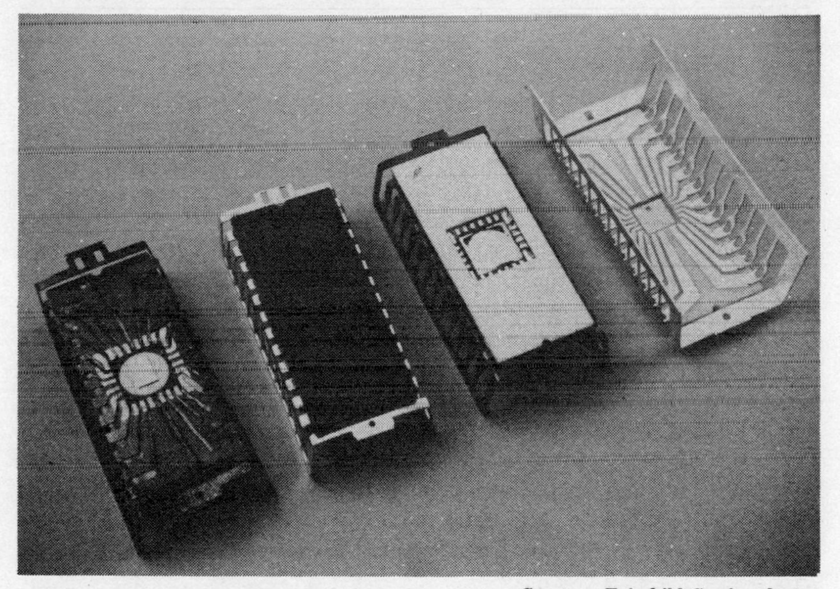

Courtesy Fairchild Semiconductor

Fig. 2-17. Dual in-line IC packages.

Once the circuits are affixed to the package, small wires are connected between the aluminum bonding pads on the integrated circuit and the lead posts of the package. The wires, which have a diameter roughly that of a human hair, are affixed either through the use of heat and pressure or through the use of ultrasonic vibration and pressure. This process of attaching the connecting wires to the aluminum pads is called *bonding*.

Fabrication of Integrated-Circuit Resistors and Capacitors

Resistors and capacitors are required, in addition to diodes and transistors, to make an integrated circuit. These devices are made concurrent with the transistor fabrication discussed previously.

The structure of an integrated-circuit resistor is shown in Fig. 2-18. The p-type diffusion used to fabricate the resistor is

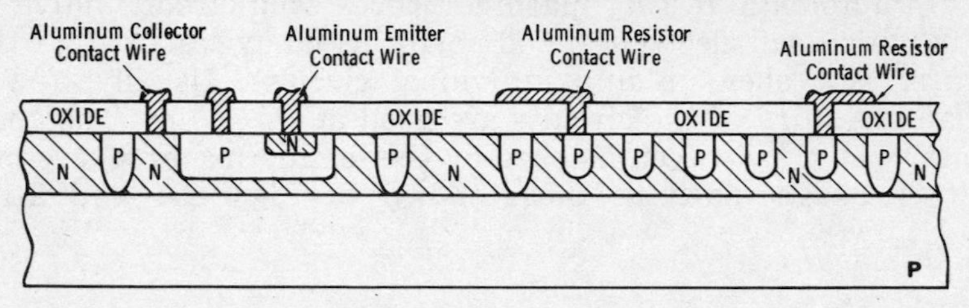

(A) Cross-sectional view.

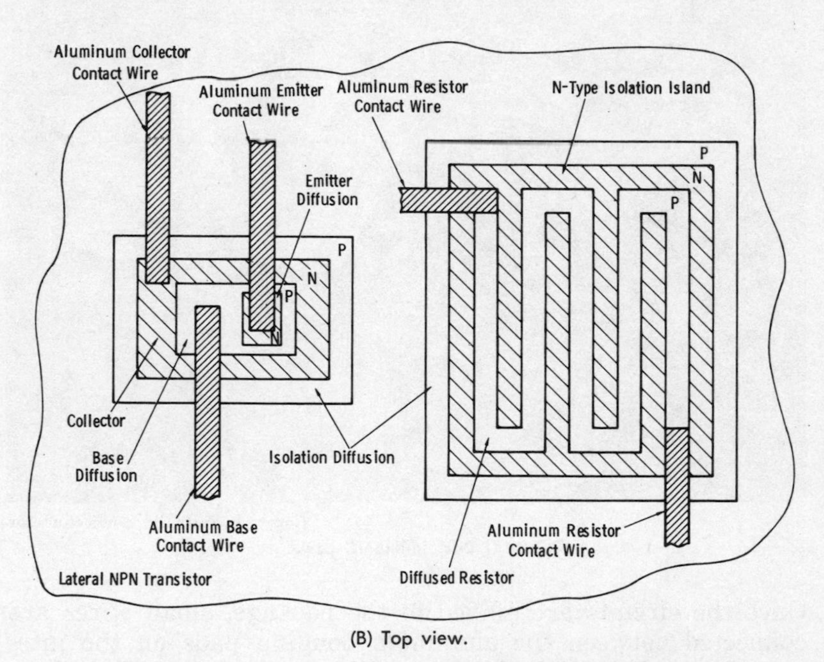

(B) Top view.

Fig. 2-18. Monolithic lateral npn transistor and diffused resistor.

the same diffusion that is used as a base diffusion in the npn transistors. All semiconductors have a resistance associated with them, which can be large if the resistor is made long with a small cross section. A typical resistor, when viewed from the top of the wafer, is long and narrow, and meanders for denser packing. Note that the n-type island in which the resistor is embedded must be connected to the highest voltage in the circuit to keep the resistor-island junction reverse biased, thereby constraining the resistor current to the confines of the resistor material.

Integrated-circuit capacitors can be fabricated in several ways. One common technique is to use the capacitance associated with the junction of a reverse-biased diode. The diffused resistor just discussed has a significant capacitance between it and the n-type island. This capacitance is dependent on the reverse voltage. (This variation of capacitance with voltage is the principle of voltage-variable capacitors, called *varactors*.)

Another technique used to build capacitors is to leave an aluminum plate on top of the silicon oxide above a diffused

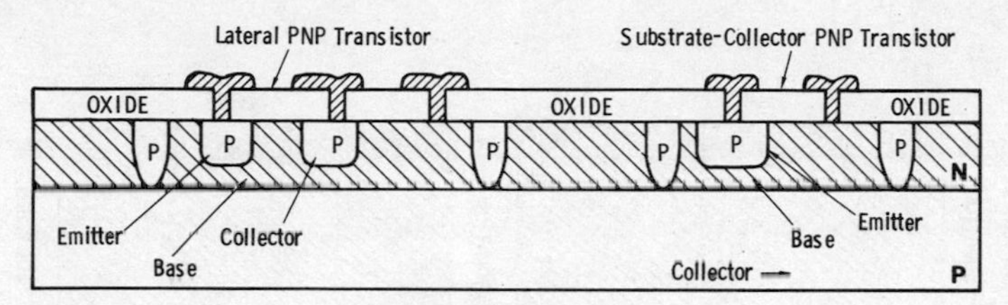

(A) Cross-sectional view.

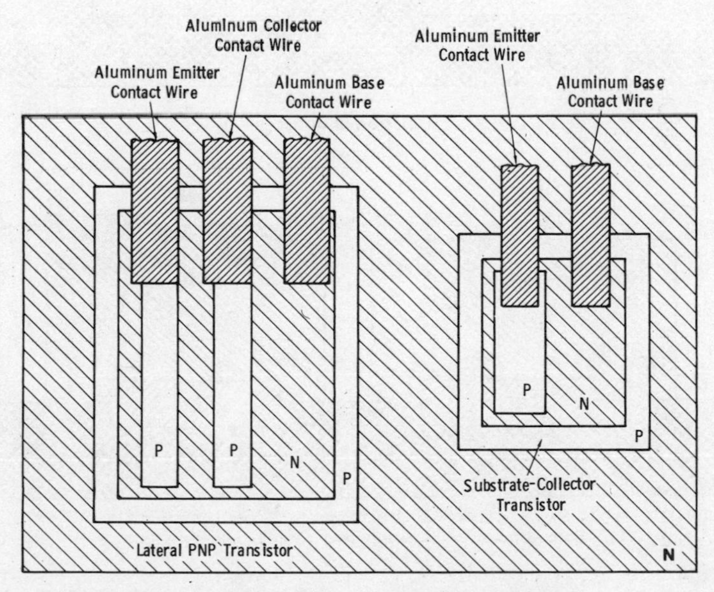

Fig. 2-19. Monolithic pnp transistor structures.

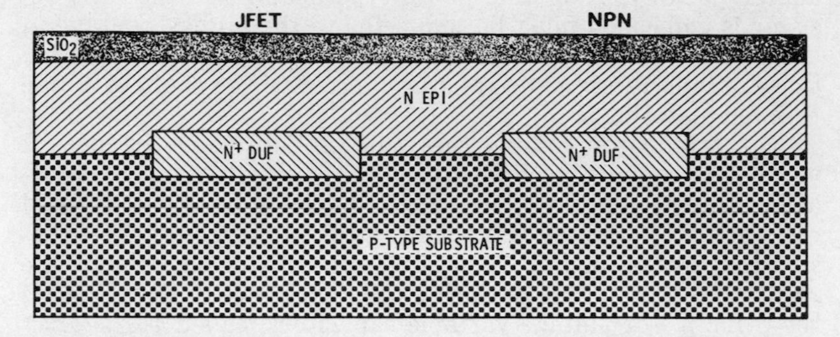

(A) Starting substrate with n+ diffusion and n-type epitaxial layer.

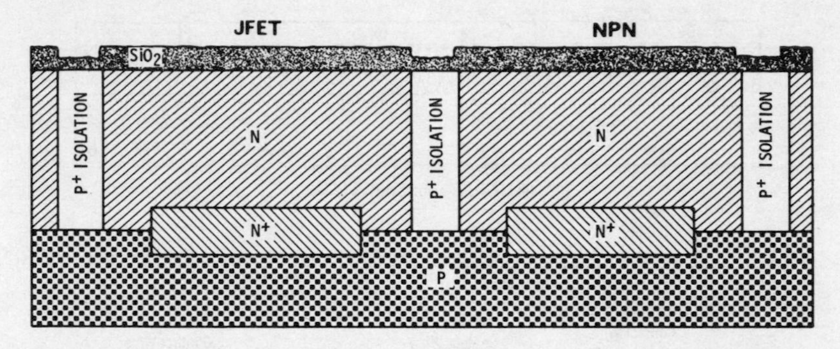

(B) Isolation diffusion.

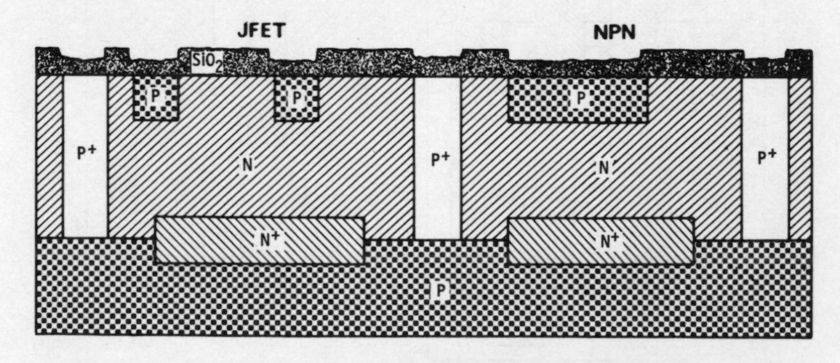

(C) Formation of JFET drain and source, and npn base.

Fig. 2-20. The BIFET

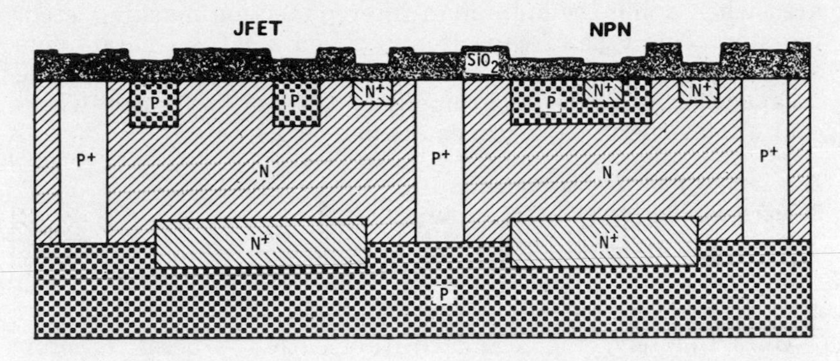

(D) Formation of npn emitter and collector, and JFET gate.

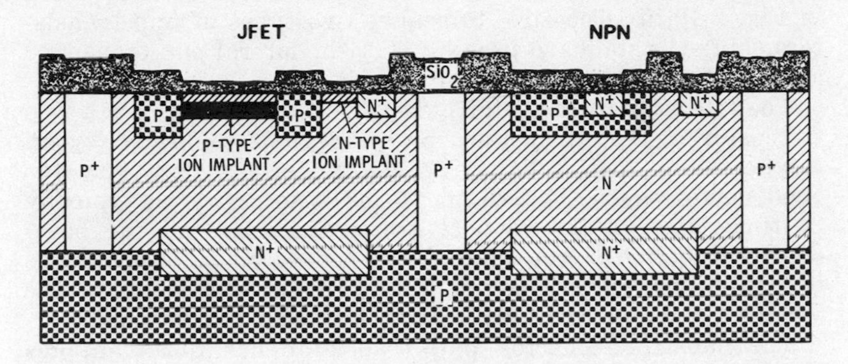

(E) JFET gate implants.

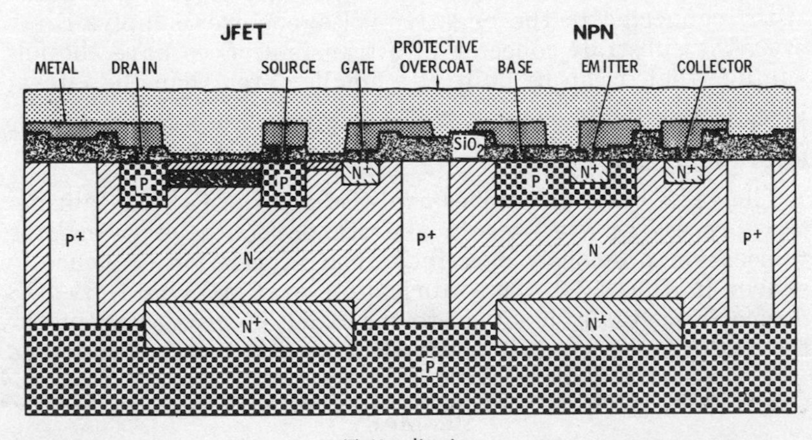

(F) Metalization.

Courtesy Texas Instruments, Inc.

fabrication process.

area when doing the aluminum interconnection masking/etching step. This type of capacitor is called a MOS (metal-oxide semiconductor) capacitor. Since the MOS capacitor can have a capacitance that is independent of voltage, it is used wherever variations in capacitance cannot be tolerated.

The Lateral PNP Transistor and Substrate-Collector PNP Transistor

Although it is possible to build almost anything with only npn transistors, it sometimes becomes very attractive to use a simpler circuit that uses a pnp transistor.

Unfortunately, the standard fabrication process is npn-transistor oriented. Additional processing steps are expensive and compromise the quality of the npn transistor. As shown in Fig. 2-19, it is possible to realize two types of pnp transistors using the standard process: (1) the lateral pnp transistor and (2) the substrate-collector transistor.

The lateral pnp transistor meets the requirements for a pnp transistor in that it has two p-type regions separated by an n-type region, but it lacks the characteristics necessary for good performance. The width of the base region is typically 20 times that of the vertical npn transistor, and the base-emitter junction is not optimally doped as p^+ as it should be to make a good p^+n junction. These factors contribute to the low gain and small bandwidth of the lateral pnp transistor.

The substrate-collector pnp transistor has disadvantages similar to the lateral pnp transistor and, in addition, it may only be used in circuits that have the collector of a pnp transistor connected to the negative-voltage power supply. However, the substrate-collector pnp transistor does have the advantage that it can be built in a smaller area than the lateral pnp transistor.

BIFET Fabrication

The same basic techniques are used for the BIFET fabrication process as are used in the standard bipolar transistor fabrication previously described. The fabrication sequence is shown in Fig. 2-20. The fabrication of CMOS and BIMOS devices is virtually identical to the BIFET process except for the addition of a metal electrode and thin insulating oxide.

SUMMARY

The material presented in this chapter provides the background necessary for the understanding of the circuitry used in IC op amps. The lateral pnp transistor, for example, is frequently used in the op-amp output stages discussed in Chapter 3. For another example, the MOS capacitor is commonly used for internal frequency compensation, as explained in Chapter 4.

CHAPTER 3

Monolithic Op-Amp Circuitry

Since the advent of monolithic integrated circuitry, several common circuits have evolved. The circuits used in monolithic IC op amps may be put into one of the following five classes: (1) amplifiers, (2) current sources, (3) voltage references, (4) level shifters, and (5) output stages. This chapter will cover the design rules that overcome the shortcomings and exploit the advantages of integrated circuits, and show how these rules lead to effective designs in each of the circuit classes. The chapter will culminate in an example of how these circuits may be joined to form an IC operational amplifier.

DESIGN RULES FOR INTEGRATED-CIRCUIT OPERATIONAL AMPLIFIERS

A set of "rules" for IC design has evolved out of the capabilities of silicon monolithic-fabrication technology. The following constitute the more significant rules:

Minimization of the Number of Package Leads

One of the major production costs of integrated circuits is the attaching of leads to the silicon chip. To keep costs down, as well as to improve circuit performance by reducing package-induced parasitic capacitances, it is important to keep the amount of external circuitry, and hence the number of leads, to a minimum.

No Integrated Inductors Allowed

Since integrated circuits are built in one plane, it is very difficult and expensive to fabricate inductors as part of the integrated circuit. Only external inductors may be used, and then only when mandatory.

Minimization of the Total Capacitance of Integrated Capacitors

As disclosed in the last chapter, capacitors may be fabricated on the silicon wafer in a standard monolithic-fabrication process. Since capacitance is proportional to the area of the plates of a capacitor, the total surface area taken up by capacitors in an integrated circuit is proportional to the total capacitance used in the circuit design. The area taken up by an integrated circuit can greatly affect the cost of that integrated circuit. Since capacitors, except for very small ones, can take up a large area in comparison to the area of a diode or transistor, the total capacitance used in IC design should be minimized.

Minimization of the Total Resistance of Integrated Resistors

As discussed in the previous chapter, IC resistors are diffused into the silicon wafer at the same time as the bases of the vertical npn transistors. Even though some area savings may be made over capacitors by being able to put all resistors in the same n-type isolation island, resistors have to be long to achieve a large resistance. Since a medium-large resistor (100,000 ohms, for instance) is much larger than a transistor or a diode, the total resistance of integrated resistors should be minimized.

Maximization of the Use of the Matching of Components

The predictability of the values of resistance, capacitance, and current gain in IC components is poor. Resistance may vary by more than 20%. Current gain may vary by more than a factor of 4. These numbers are the variations from the desired values. On a single silicon wafer, however, the matching between similar components is much better than this because all the components on the wafer are fabricated at the same time and receive identical treatment. A good IC design will exploit this excellent "within-the-wafer matching" by making the design, whenever possible, dependent largely on the ratio of values of components and not on their absolute values. This is an important reason for the frequent use of differential circuits in IC design. Differential circuits will be discussed in detail later in the chapter.

Maximization of the Substitution of Transistors and Diodes for Larger-Area Components

Diodes and transistors are typically small structures by comparison to the other components on the silicon chip. Transistors that have collectors in common with other transistors, and diodes that have n-type sides in common with other diodes or transistors can be made even smaller by sharing the same n-type island. Before being integrated, a design is checked to see if it is possible to change the circuit to use diodes or transistors in place of larger-area capacitors and resistors. As shown in Fig. 3-1, it may save area and improve performance to replace a large resistor with two smaller resistors and two transistors. In some applications, the junction field-effect transistor can be used almost as a direct replacement for high values of resistance, as shown in Fig. 3-2.

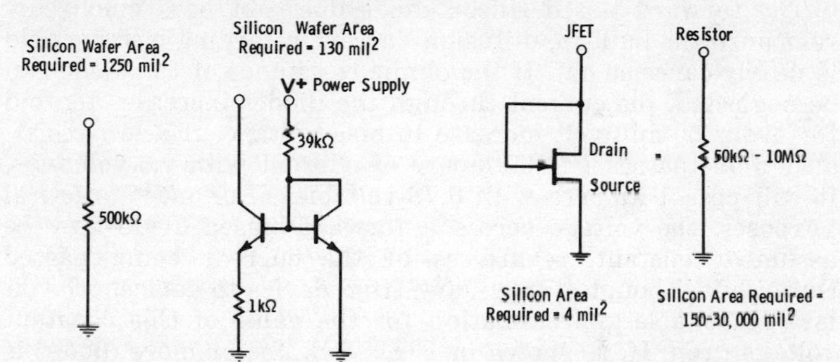

Fig. 3-1. Area comparison for circuit alternatives.

Fig. 3-2. Circuit connection of JFET to simulate a large resistor.

Maximization of the Use of Built-In Voltage References

A voltage reference is a device that can be used to create a voltage which, among other things, may be considered constant and independent of power-supply voltage. In integrated circuits, there are two common sources of voltage references: (1) the breakdown of reverse-biased diodes and (2) the voltage drop across forward-biased diodes.

The reverse-biased diode, discussed in Chapter 2, is shown in a typical circuit in Fig. 3-3A. In an integrated circuit, the diode would be the base-emitter junction of a vertical npn transistor. For typical ICs, this junction breaks down at around 5 volts. The other junction, the vertical npn transistor collector-base junction, built during the standard fabrication procedure, has too high a breakdown voltage to be useful in integrated-circuit design.

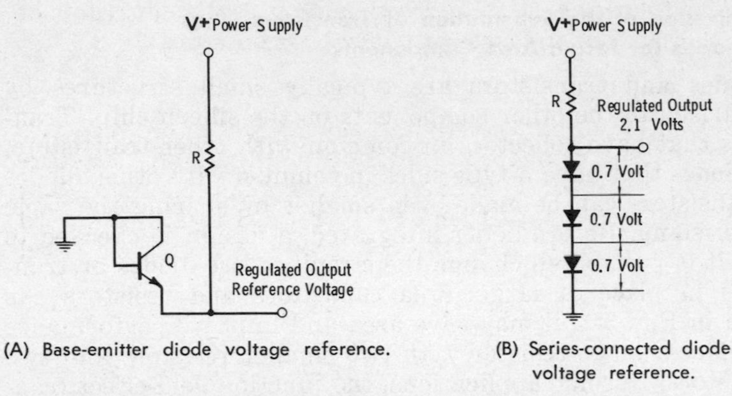

(A) Base-emitter diode voltage reference.

(B) Series-connected diode voltage reference.

Fig. 3-3. Constant-voltage reference sources.

The forward-biased silicon diode does not pass much current until the built-in, diffusion-current retarding electric field is nearly canceled out. If the ohmic resistance of the diode can be neglected, the current through the diodes increases tenfold for every 60-millivolt increase in bias voltage. If a low-resistance diode passes 1 milliampere of current with 0.6-volt bias, it will pass 1 ampere with 0.78-volt bias. For most practical purposes, the voltage across a forward-biased diode may be assumed constant, regardless of the current being passed through it. Though it may vary from device to device, 0.7 volt is a reasonable approximation for the value of this constant voltage drop. If, as shown in Fig. 3-3B, one or more diodes is put in series with resistance R, a very good and convenient reference voltage may be obtained.

A voltage reference creates noise, regardless of the type of device used to generate it. The breakdown of the reverse-biased diode inherently has substantially more noise output than the forward-biased diode. Hence, the breakdown of reverse-biased diodes should not be utilized in a signal path or where the minimization of noise is important.

Integrating the reference-voltage device on the same silicon chip reduces the number of package leads and results in a less-expensive unit. Having the reference device on the same silicon chip also allows closer temperature tracking, and hence less drift.

IC OP-AMP CIRCUITS

Although it is impossible to show all circuits used in IC op amps, it is relatively easy to classify parts of all op-amp circuits. The following is a description of a complete group of circuit classes that are used in almost all op amps:

Differential Amplifiers

A differential amplifier amplifies the difference between two input signals. A circuit for a typical differential-amplifier stage is shown in Fig. 3-4. This particular amplifier uses standard bipolar transistors, but either JFET or MOS transistors used in the BIFET, BIMOS, and CMOS operational amplifiers could be used as well.

The current-source transistor (Q_3) is biased from the power supply. To a good degree of approximation, the current flowing in the collector (I_{CQ3}) is a constant value set by the base current (I_{BQ3}) times the current gain (β) of transistor Q_3. When both inputs are grounded, one half of I_{CQ3} will flow through transistor Q_1, and the other half will flow through transistor Q_2, if the transistors are well matched. Since one half of current I_{CQ3} is flowing through Q_1, a base-emitter current (I_{BQ1}) will flow, allowing the much larger collector-emitter current (I_{CQ1}) to flow. Both I_{BQ1} and I_{CQ1} add to make up half of I_{CQ3}. Likewise, a base-emitter current (I_{BQ2}) and collector-emitter current (I_{CQ2}) will flow through transistor Q_2, comprising the other half of I_{CQ3}. These base currents (I_{BQ1} and I_{BQ2}) are called the *input-bias currents*.

If, as shown in Fig. 3-5, both inputs 1 and 2 are raised by ΔV (called a *common-mode input signal*), the output remains unchanged because both inputs were treated identically and

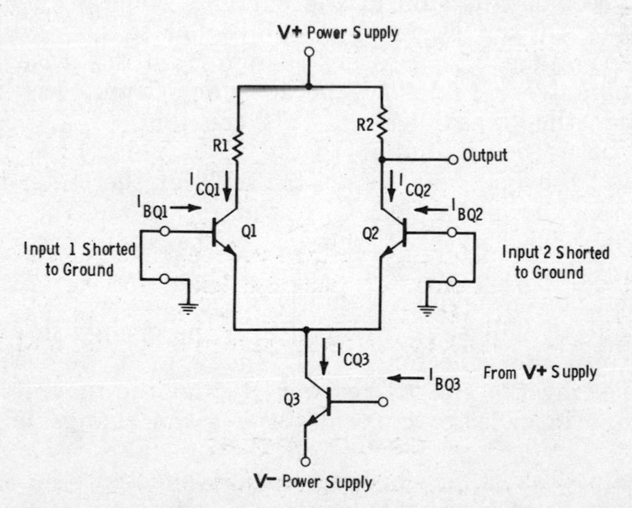

Fig. 3-4. A typical differential-amplifier circuit with inputs grounded.

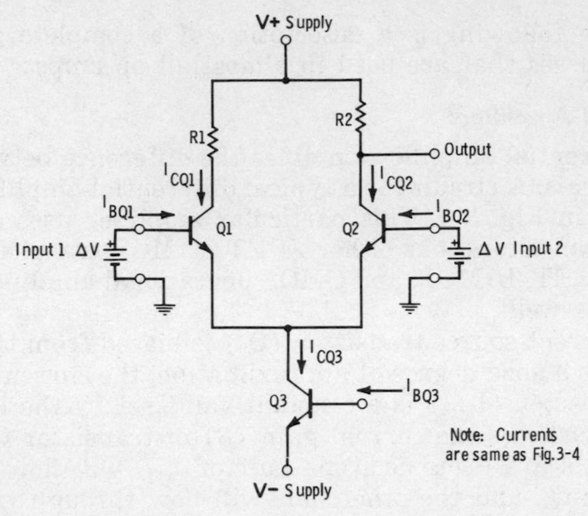

Fig. 3-5. A typical differential amplifier with a common-mode input signal.

the circuit is symmetrical. The only change that does occur in the circuit is that V_{CQ3} increases by ΔV. Note that I_{CQ3} does not change because the collector-emitter current is dependent only on the base-emitter current, and not on the collector-emitter voltage.

If, as shown in Fig. 3-6, input 1 is raised by ΔV and input 2 is lowered by ΔV (called a *differential-mode input signal*), the current flowing through Q_1 will increase, while the current flowing through Q_2 will decrease by the same amount, because the sum of the currents coming out of the emitters of Q_1 and Q_2 must remain equal to I_{CQ3}, which is constant. Voltage V_{CQ3} does not change from the case where both inputs were grounded, because one input was raised while the other was lowered. This constant (V_{CQ3}) can be thought of as being similar to the pivot point of a teeter-totter, and the input signals as the ends of the teeter-totter.

Since the current through Q_1 has increased and the current through Q_2 has decreased, there is a change in the voltage drop across the load resistors. Since the output of this amplifier is the power-supply voltage less the voltage drop across R_2, the output will have a response to the change in current flowing into the collector of Q_2. The gain of the stage is determined by the size of resistor R_2 and the magnitude of the change in collector current for a given change in input voltage.

Assuming that I_{CQ3} remains constant and that the components are well matched, it is clear that this differential-amplifier

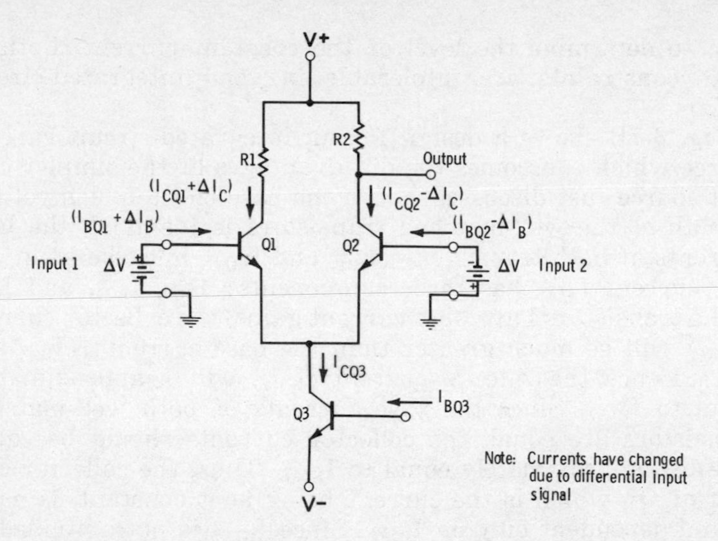

Fig. 3-6. A typical differential amplifier with a differential-mode input signal.

circuit will amplify a differential-mode input signal but will not amplify a common-mode input signal. In practical circuits it is impossible to match components perfectly and to keep I_{CQ3} absolutely constant as a function of collector-emitter voltage. These nonideal factors cause the amplifier to have some output signal in response to a common-mode signal as well as to a differential-mode signal. Typically, however, the amplifier amplifies a differential-mode signal 10,000 times more than it amplifies an equivalent common-mode signal. Yet this performance can only be achieved with well-matched transistors, such as those that were fabricated at the same time, side by side, on the same silicon wafer.

Constant-Current Sources

Often in operational-amplifier design there is a need to maintain a constant current flowing. An example is I_{CQ3}, discussed in the previous section. Usually, integrated-circuit constant-current sources are constant-current *sinks*. They are designed using a simple transistor model that is based on the assumption that the collector-emitter current of a transistor is equal to the base-emitter current times the current gain of the transistor, and that the current gain is virtually independent of the collector-emitter voltage.

The simple current source referred to in the previous section is shown in Fig. 3-7A. This circuit requires large values of resistance and depends on the value of the transistor current gain to determine the level of the constant current. Both of these constraints are intolerable in good integrated-circuit design.

Fig. 3-7B shows a design for an integrated-circuit current source which overcomes the disadvantages of the simpler current source just discussed. Since the base-emitter bias voltage of both of the well-matched transistors is identical, the base currents of both transistors (I_{BQ3} and I_{BQ4}) must be identical. The current I_{BIAS} has three components: I_{BQ3}, I_{BQ4}, and I_{CQ4}. If the transistors have high current gains, the collector current (I_{CQ4}) will be much greater than the base currents (I_{BQ3} and I_{BQ4}). Hence, the collector current (I_{CQ4}) will be approximately equal to I_{BIAS}. Since the base currents of both well-matched transistors are equal, the collector currents should be equal, and also approximately equal to I_{BIAS}. Thus, the collector current of Q_3, which is the current being kept constant, is equal to and dependent only on I_{BIAS}. Since I_{BIAS} is not amplified, it is a larger current than supplied in the simpler current source shown in Fig. 3-7A. Hence, for a given power-supply voltage a smaller, space-conserving resistor may be used than in the simpler circuit. In applications where a very small constant current is desired, $I_{CONSTANT}$ may be made less than I_{BIAS} by placing a resistor in series with the emitter of Q_3, as shown in Fig 3-7C. This resistor serves to decrease bias across the base-emitter junction, which decreases the base-emitter current. This in turn decreases the collector current, which is the current being controlled.

The goal of these designs is to utilize the matching of transistors to achieve a stable, current-gain independent, current source.

Voltage References

There are two types of voltage references used in operational-amplifier design: (1) the forward-biased diode and (2) the reverse-biased diode. The voltage drop across the forward-biased diode is the most popularly used voltage reference in operational amplifiers.

The current source shown in Fig. 3-7B, and discussed in the last section, is biased by the base-emitter junction voltage of transistor Q_4 being applied to the base-emitter junction of Q_3. The base-emitter junction of Q_4 is acting like a special type of forward-biased diode. The application of this built-in reference voltage results in a very high-performance circuit, because any changes due to temperature, for instance, cancel out if the transistors are well matched.

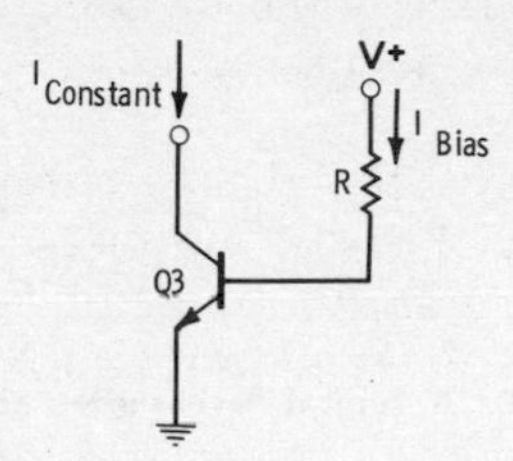

(A) A simple constant-current source.

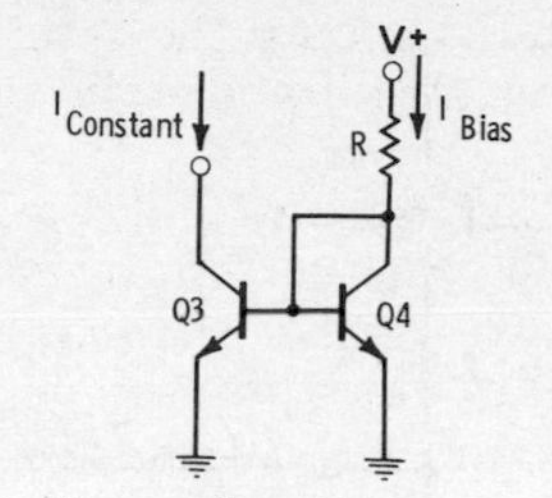

(B) An improved constant-current source.

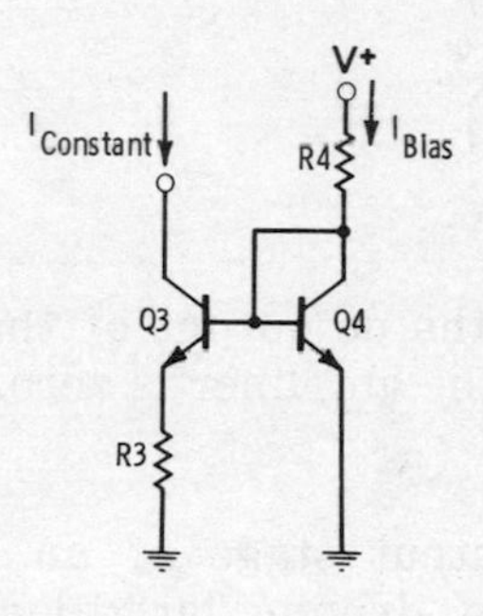

(C) An improved constant-current source for small currents.

Fig. 3-7. Constant-current sources.

Level Shifters

In a typical operational amplifier there often are stages of differential amplifiers, followed by a level shifter which drives an output stage. There is a need for level shifters because there is a dc offset in the output of a differential amplifier, such as shown in Fig. 3-4 for the case of no input.

The exact type of level shifting required depends on the exact circuit used in the preceding amplifier stages and on the output stage it drives. A typical level-shifting circuit, which provides no voltage gain, is shown in Fig. 3-8.

Transistor Q_6 acts as a current source which maintains V_{CQ6} constant. The voltage drop across resistor R_5, using Ohm's law, is:

$$V_{R5} = I_{CQ6} \times R_5$$

This voltage drop is a constant since I_{CQ6} and R_5 are constant. The voltage drop across the base-emitter junction of Q_5 (V_{BQ5}) is simply a forward-biased diode voltage drop (approximately 0.7 volt), which is also constant. Since the output voltage is just the input voltage less the drop across the base-emitter junction of Q_5 and the drop across R_5, the output of this stage can be thought of as the input voltage shifted in dc level by the sum of the two voltage drops. These drops can be designed

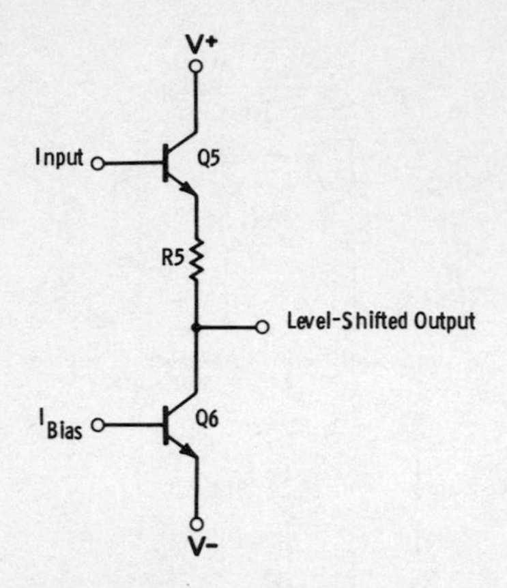

Fig. 3-8. A typical level-shifter circuit.

to make the dc output of the entire amplifier zero when the input to the amplifier is zero.

Output Stages

The output stage of an operational amplifier should be capable of driving large loads. In addition, it is sometimes desirable that the amplifier be capable of providing large output voltage swings as close as possible to the power supplies. The CMOS operational amplifiers include complementary MOS transistors in the output stage which provide maximum output voltages within a tenth of a volt of the power supplies. The current output of the MOS output amplifier stages, however, is less than the bipolar npn/pnp complementary transistor stages.

A typical npn/pnp complementary transistor output stage is shown in Fig. 3-9. The operation of this circuit may be understood by realizing that the output voltage is different from the base voltages of output transistors Q_7 and Q_8 by just a forward-biased, base-emitter diode voltage drop (0.7 volt). Thus, any signal appearing at the connection of diodes D_1 and D_2 will appear at the output. These diodes are incorporated into the level-shifter circuit of Fig. 3-8 so that the output voltage is just equal to the input voltage shifted in dc level. Although the output stage has no voltage gain, it does have power gain and is able to drive large loads.

This configuration is capable of delivering large currents to a load, yet has small standby currents flowing if no input signal is applied. If a large (low-resistance) load is placed on the output, the output voltage will not drop significantly. This is because a small drop in the output voltage tends to increase and decrease the appropriate output-transistor base-emitter

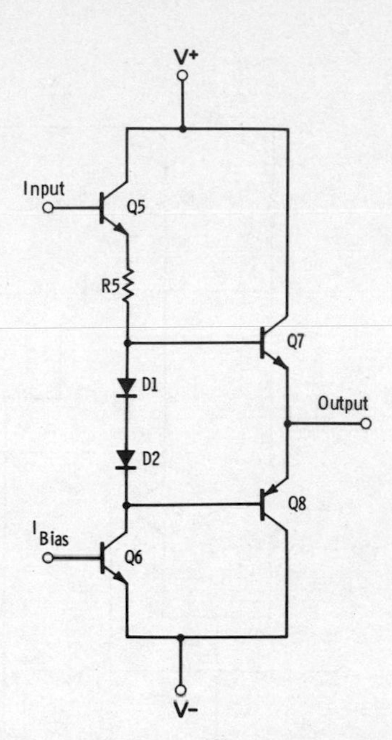

Fig. 3-9. A typical npn/pnp complementary output stage.

biases in such a direction as to supply more current to the load. This circuit may be thought of as incorporating a form of negative feedback to increase the load-driving capability of the output stage.

Since the output stage of an integrated-circuit operational amplifier can heat up, it is often placed as far away from the differential input stage as possible on the silicon chip, to reduce drifts in the output voltage due to temperature effects.

A SIMPLE INTEGRATED-CIRCUIT OPERATIONAL AMPLIFIER

Thus far, this chapter has covered the design rules of integrated circuits and a sampling of the circuits used in operational amplifiers. It can be seen that the circuits exemplify the philosophy behind the rules.

The circuits described in this chapter have been chosen to be representative of those used in operational amplifiers. A complete operational amplifier that utilizes these circuits is shown in Fig. 3-10. The operation of this circuit should be apparent from the previous descriptions. (Notice that voltage reference Q_4 is shared by Q_3, Q_6 and Q_9, and that resistors R_3 and R_6 are used to reduce the currents flowing through Q_3 and Q_9 to their optimal values. Also notice that the input dif-

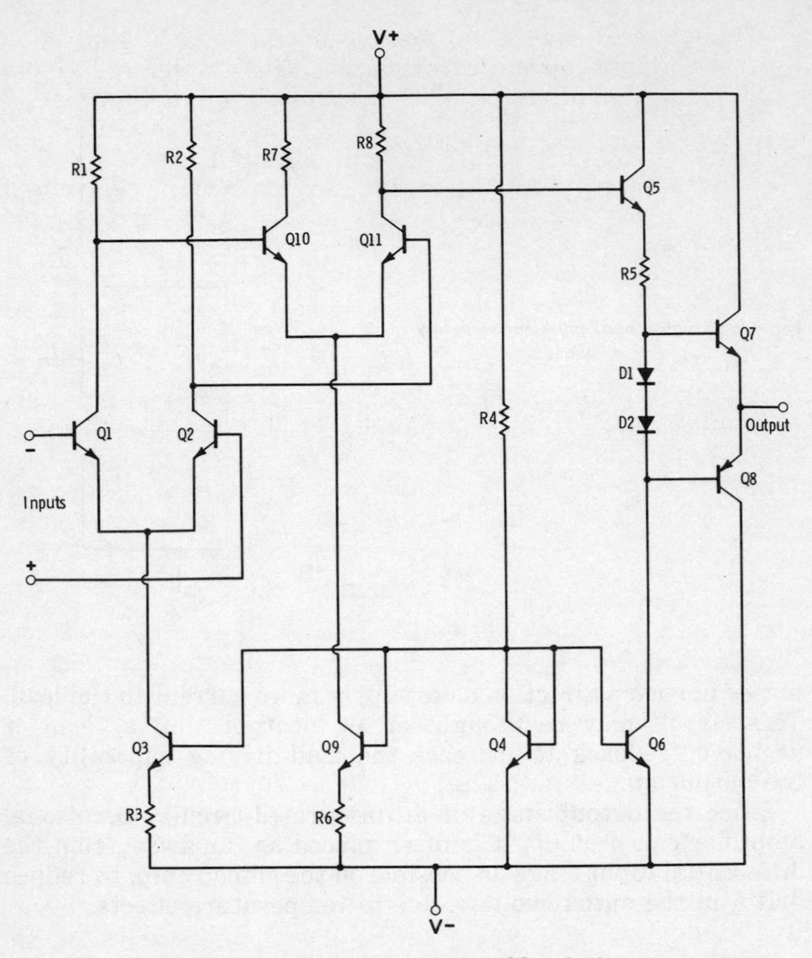

Fig. 3-10. A simple operational-amplifier circuit.

ferential amplifier has outputs taken from both collector resistors R_1 and R_2.)

THE 741 OPERATIONAL AMPLIFIER

Commercially manufactured operational amplifiers can be analyzed in terms of the circuit classes discussed in this chapter. The actual circuits used in the commercially fabricated devices may be simpler or far more complex than the circuits that have been discussed. The same general design philosophy of integrated-circuit technology is used in all the circuits, however. In addition to being a widely used, high-performance, general-purpose operational amplifier, the 741, shown in Fig.

3-11, is a good example of an op amp that can be analyzed in terms of the concepts introduced in this chapter; yet it is a fairly different circuit from the simple circuit of Fig. 3-10.

Differential-Amplifier Input Stage

The differential-amplifier input stage of the 741 is shown in Fig. 3-12. Transistor Q_{11} acts as a voltage reference for current-source transistor Q_{10}, and transistor Q_8 acts as a voltage reference for current-source transistor Q_9. It is easy to identify this type of voltage reference from the connection between the collector and base terminal of the transistor.

Since the two current sources are connected together, the currents flowing through them will be approximately equal, assuming that the collector-emitter current flowing through them (I) is much larger than the current going into the bases of transistors Q_3 and Q_4. As discussed previously in the constant-current section, the current flowing through the voltage-reference diode is approximately equal to this constant current I. The other voltage reference (Q_{11}) has a current I_{BIAS} (which is greater than I due to resistor R_4) flowing through it. Since I_{BIAS} is determined by the power supply voltage, it may be considered constant. Note that I_{BIAS} determines all the currents flowing through the differential-amplifier input stage by adjusting the base current flowing in transistors Q_3 and Q_4, which in turn equalizes the current flowing through current-source transistors Q_9 and Q_{10}.

If both inputs are grounded and the transistors are well matched, the currents will be distributed as shown in Fig. 3-12. Since the current gain of the transistors is high, the base currents are small in comparison to the collector and emitter currents and may be ignored when analyzing the collector and emitter currents.

This differential amplifier will ideally have no response in the output, due to a common-mode input signal. That is, raising or lowering both inputs will result in no change in the currents flowing through transistors Q_1 and Q_2 because both inputs are treated equally.

Analyzing the response of the input stage to a differential-mode input signal is a bit more complex, as shown in Fig. 3-13. Transistors Q_5 and Q_6 must always have the same collector-emitter current flowing through them since they receive, by symmetry, the same base-emitter bias, which comes from transistor Q_7. Transistor Q_7 is in turn controlled by the collector voltage on transistor Q_5. Thus, the collector-emitter current flowing through transistor Q_6 is always equal to the collector-

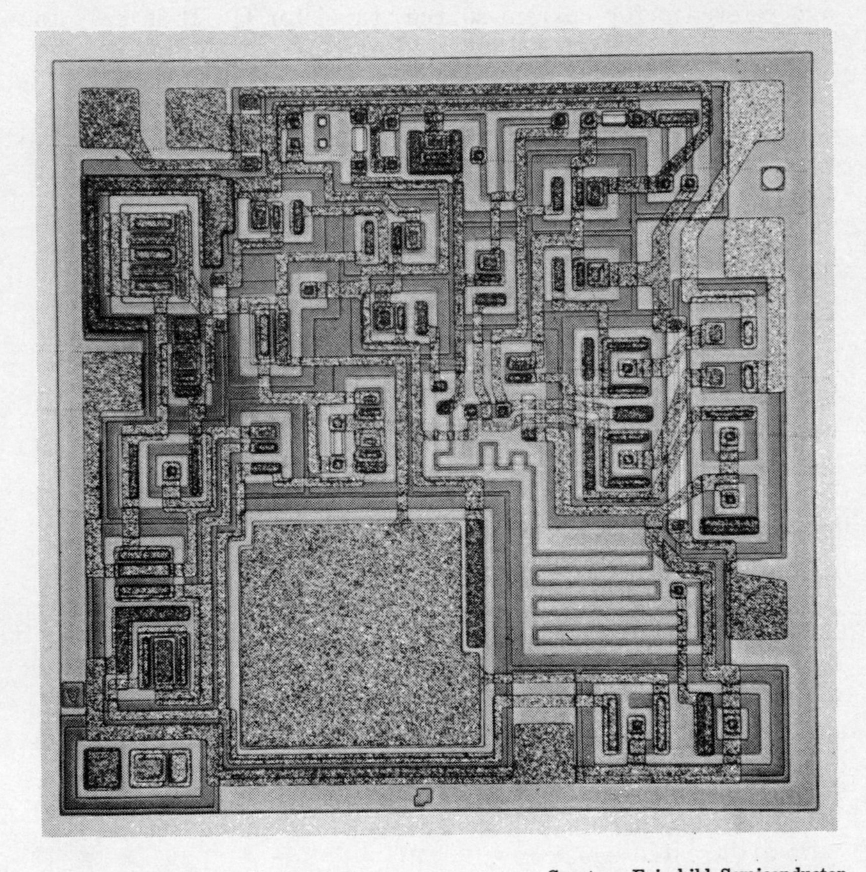

Courtesy Fairchild Semiconductor

(A) Photomicrograph of the 741 op-amp silicon chip.

Fig. 3-11. The 741

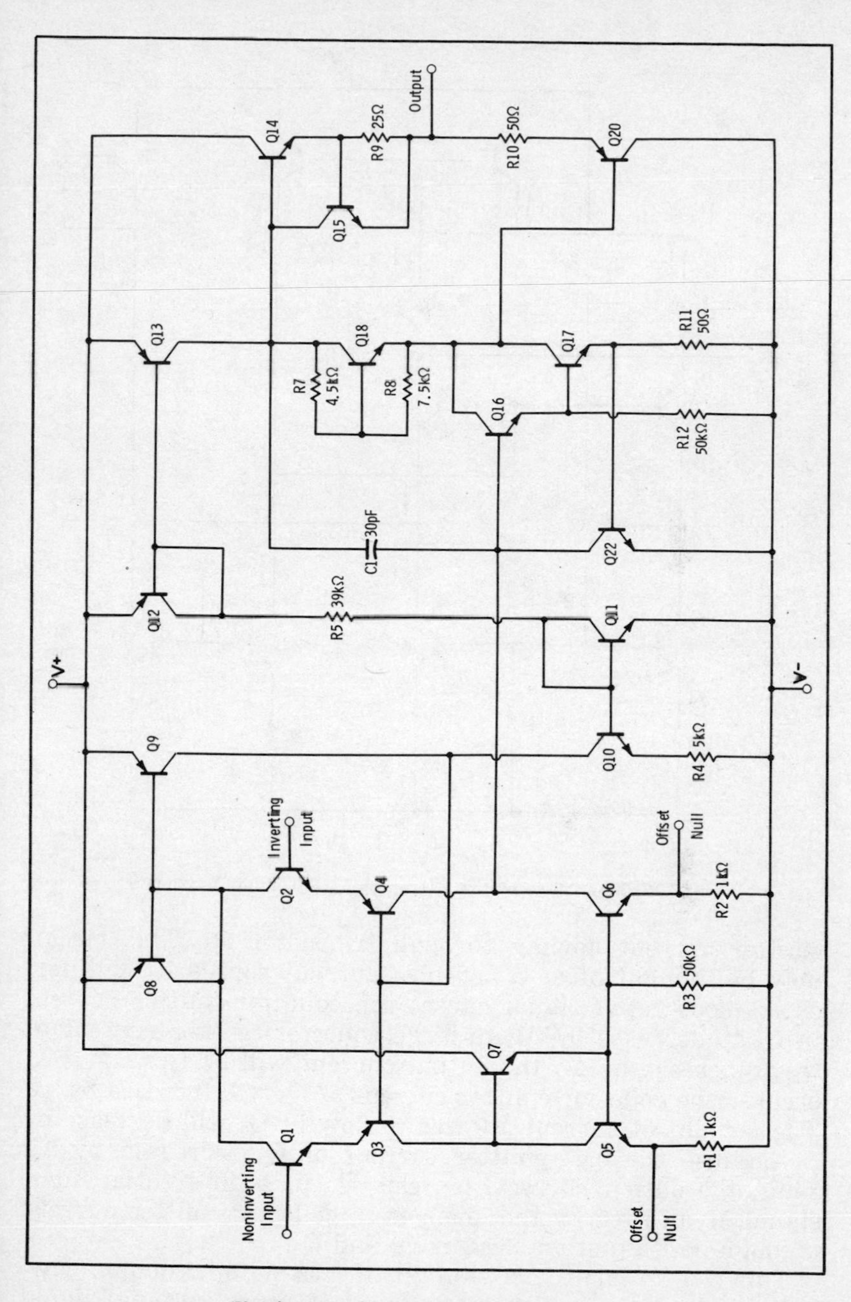

(B) Schematic diagram of the 741 op amp.

operational amplifier.

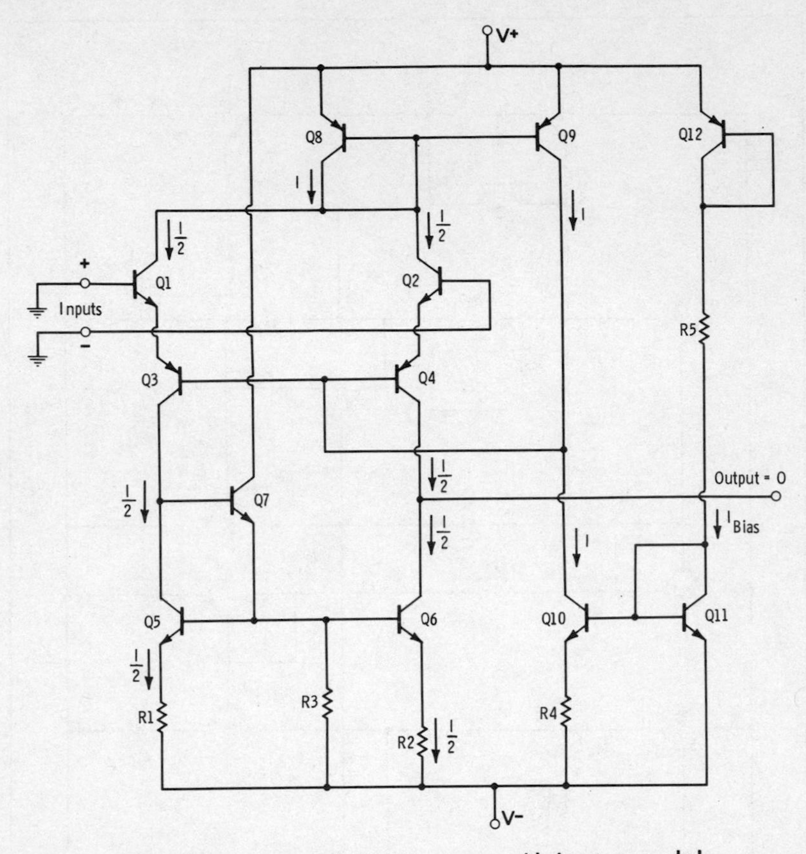

Fig. 3-12. Input stage of the 741 op amp with inputs grounded.

emitter current flowing through transistor Q_5. This circuit may be thought of as a modified current source. If a differential-mode input signal causes the collector-emitter current in Q_1 to decrease by ΔI, and the collector-emitter current of Q_2 to increase by ΔI, the output current will be two times ΔI because the collector-emitter current of Q_4 will increase by ΔI. The amount of current allowed to flow in Q_6 will decrease by ΔI because the base-emitter current of Q_5 decreased by ΔI. Thus, the output current, in response to a differential input signal, is affected by the changes in collector-emitter current through both input transistors Q_1 and Q_2.

This circuit has the advantage of a large, maximum-allowable differential input signal because a differential input signal would have to "break down" two reverse-biased emitter-base junctions in comparison with one junction of the previously

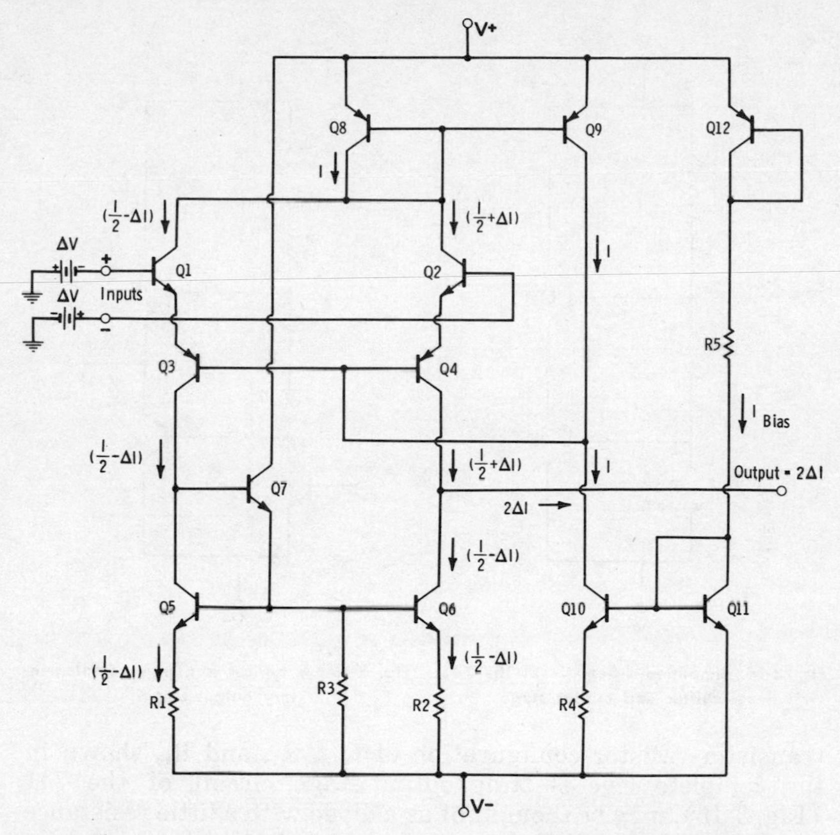

Fig. 3-13. Input stage of the 741 op amp with differential input signal.

discussed differential amplifiers. The additional junction is the base-emitter of a lateral pnp transistor. This junction has a large reverse-biased breakdown voltage. The combination of the two junctions in series then increases the maximum nondestructive input signal many times. This improvement is very important, especially in integrator and comparator applications, and makes the input-stage transistors virtually indestructible.

Level Shifter and Output Stage

As seen in the simplified circuit shown in Fig. 3-14, the level shifter and output stage of the 741 are very similar to the circuit of Fig. 3-9, previously discussed in this chapter and shown again in Fig. 3-15. Due to the input stage configuration, the level shifting of the 741 is in the opposite direction (in the direction of higher voltages) to that of the previous circuit. The

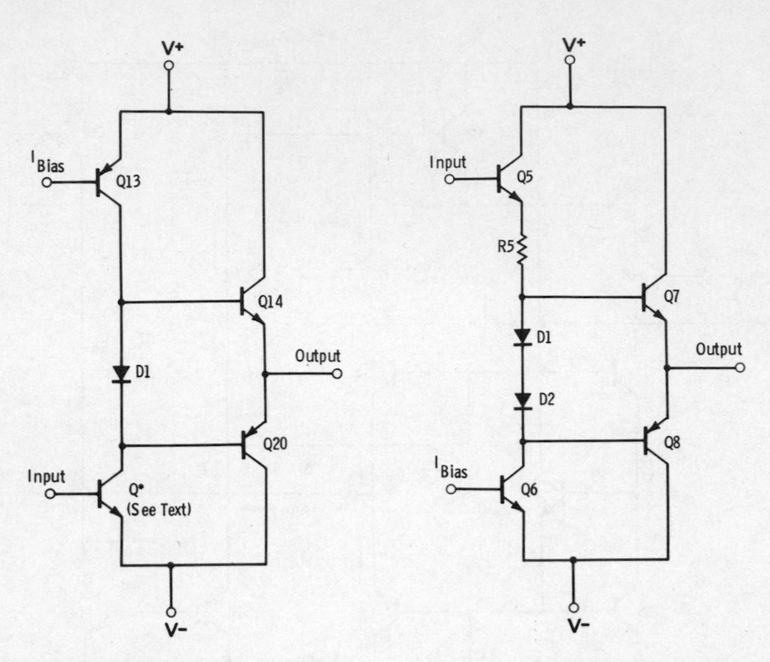

Fig. 3-14. Simplified schematic of the 741 level shifter and output stage.

Fig. 3-15. A typical npn/pnp complementary output stage.

transistor-resistor configuration of Q_{18}, R_7, and R_8, shown in the complete level-shifting/output-stage circuit of the 741 (Fig. 3-16), may be thought of as a diode with a little resistance added to increase the voltage drop across it. This voltage drop is a little more than the drop across the single diode (D_1) of Fig. 3-14, and a little less than the drop across the two diodes (D_1 and D_2) of Fig. 3-15. This is a compromise in biasing the output stage between distortion and power consumption. The transistor (Q*) in the simplified circuit of Fig. 3-14 represents transistors Q_{16} and Q_{17} in the complete circuit of Fig 3-16. This configuration of transistors, which is known as the Darlington configuration, simulates a very high current gain transistor by injecting the emitter current of the first transistor (Q_{16}) into the base of the second transistor (Q_{17}). Since the amplified current of the first transistor is put into the input of the second, the two transistors are roughly equivalent to a single transistor with a current gain equal to the current gain of the first transistor times the current gain of the second transistor. This type of level shifter and output stage has a signal-voltage gain, in contrast to the configuration of Fig 3-8, which has unity gain.

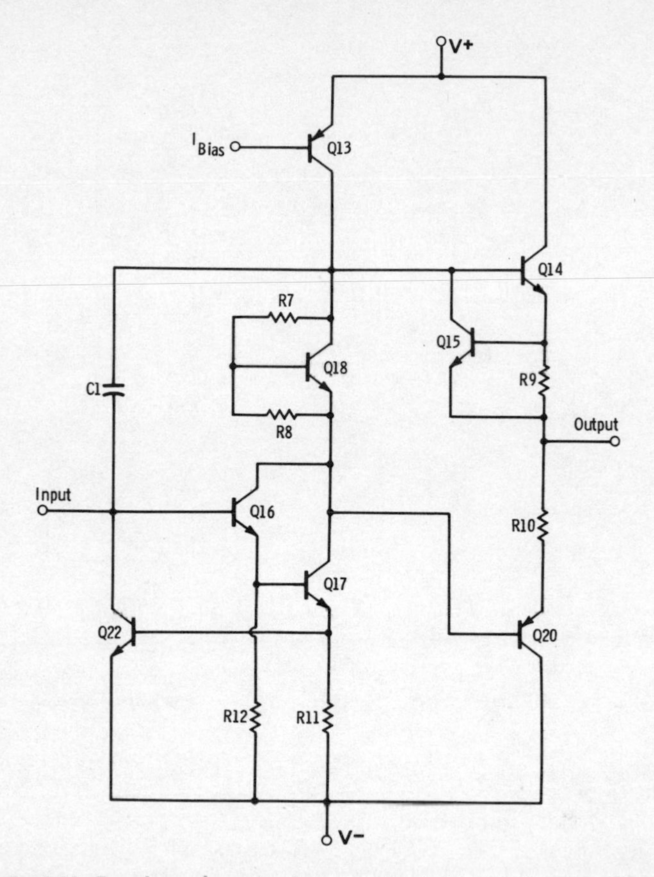

Fig. 3-16. Complete schematic of the 741 level shifter and output stage.

Transistors Q_{22} and Q_{15}, shown in the complete circuit of Fig. 3-16, act as output-current limiting transistor "switches" that turn on whenever the currents flowing through resistors R_{11} and R_9, respectively, reach their threshold voltage (approximately 0.7 volt). When turned on, these switches short out the base-emitter current of the output transistors. These protection transistors, in effect, limit the output current to some predetermined amount.

The 741 is one of the more advanced and sophisticated operational amplifiers, yet it and all other integrated-circuit operational amplifiers may be analyzed in terms of the circuit concepts and classes discussed in this chapter.

CHAPTER 4

The Integrated-Circuit Op Amp

Circuits that use operational amplifiers are often designed using the concept of the ideal op amp discussed in Chapter 1. It is often possible to obtain an operational amplifier that will behave as an ideal op amp for the application of interest. Integrated-circuit operational amplifiers do have limitations, however, and it is the purpose of this chapter to identify them, show their origin, and show how to predict their effect from a manufacturer's specification sheet.

OUTPUT OFFSET

The steady-state output of an ideal op-amp negative-feedback circuit is zero when the input is zero. In the same circuit, however, an IC op amp may have a dc output voltage, called the *output offset voltage*. If the output offset voltage is large, it may adversely affect circuit performance. Output offset voltage is due to two sources: (1) *input bias current* and (2) *input offset voltage*.

Bias-Current Offset

The input bias currents were discussed in the section on differential amplifiers in Chapter 3. They are essentially the currents that must be supplied to the two inputs of the operational amplifier to assure proper biasing of the differential input-stage transistors. Since bipolar transistors are current-amplifying devices [as opposed to unipolar transistors (FETs), which are voltage-amplifying devices], input bias currents must always be supplied, even with perfect component matching. Even FETs require some input bias current, as discussed in Chapter 2. The operational amplifiers utilizing the BIFET technology (080, 081, etc.) are useful in applications where this bias current must be small. The BIMOS and CMOS devices (3130, 3140) are useful in applications where this current should be even smaller.

The input bias current, in the simple inverter circuit shown in Fig. 4-1, flows through resistors R_1 and R_2, causing a voltage to be developed across them. This voltage appears at the input of the op amp as a differential input signal because there is no corresponding voltage appearing at the other (grounded) input terminal of the IC. It is this differential input signal that is amplified by the op amp. The output offset voltage due to input bias current in this simple inverter circuit is approximately:

$$E_{os} = I_b \times R_2$$

where,

E_{os} is the output offset voltage in volts,
I_b is the input bias current in amperes,
R_2 is in ohms.

This equation is derived assuming that the differential input signal is virtually zero for any output voltage.

A simple way to correct for output offset due to bias currents is to place a resistor (R_3) between the +input and ground, as shown in Fig. 4-2. The voltage drop across R_3 (due to bias current) must be equal to the voltage at the −input caused by bias currents flowing through R_1 and R_2. To have an output offset of zero, the value of R_3 would be:

$$R_3 = \frac{R_1 \times R_2}{R_1 + R_2}$$

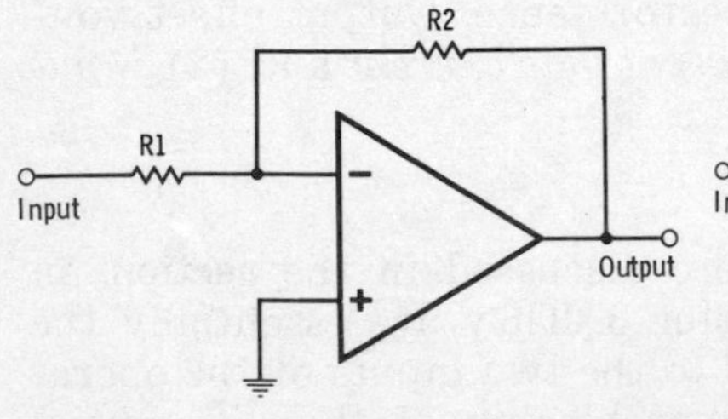

Fig. 4-1. A simple inverting amplifier.

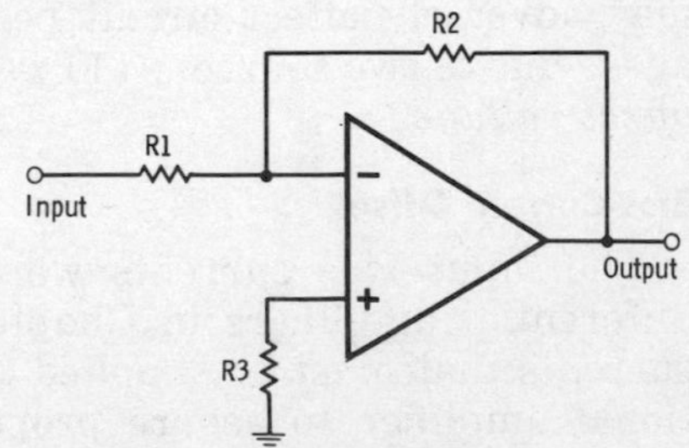

Fig. 4-2. Bias-current-balanced inverting amplifier.

assuming that the bias currents flowing into each lead are equal.

But even though the bias currents are well matched, they are not *exactly* equal. In the circuit of Fig. 4-2, there will still be an output offset voltage due to the difference between the two bias currents. The difference between the bias currents flowing into the two inputs, when the output is at zero volts, is defined as the *input offset current*. The output offset voltage due to the input offset current is approximately:

$$E_{os} = I_{os} \times R_2$$

where,

E_{os} is the output offset voltage in volts,
I_{os} is the input offset current in amperes,
R_2 is in ohms.

The addition of resistor R_3 will reduce the output offset voltage because the difference in the bias currents flowing into the op-amp inputs (i.e., the input offset current) is less than either of the bias currents. The output offset voltage using this technique may be of either polarity, depending on the internal mismatches in the op amp.

Input Offset Voltage

The *input offset voltage* is the other source of output offset. The input offset voltage, as shown in Fig. 4-3, is defined as equal to the differential input voltage that must be applied across the inputs of the op amp in order to make the output voltage zero. The input offset voltage is not zero due to slight mismatches in the internal components of the IC op amp. The FET types of IC op amps (080, 3140, etc.) typically have more input offset voltage than their bipolar counterparts (741, etc.).

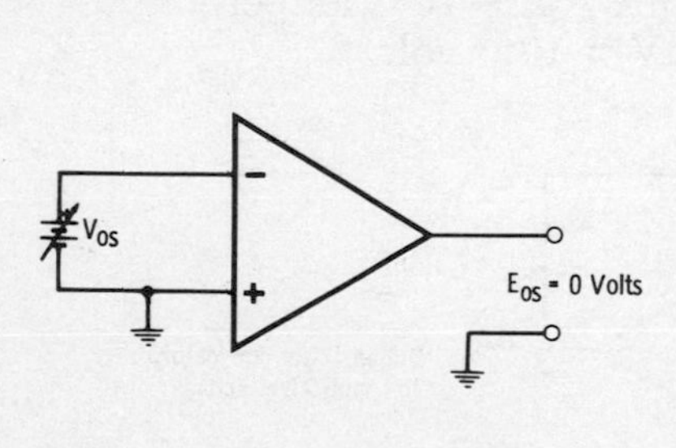

Fig. 4-3. Illustration of input offset voltage.

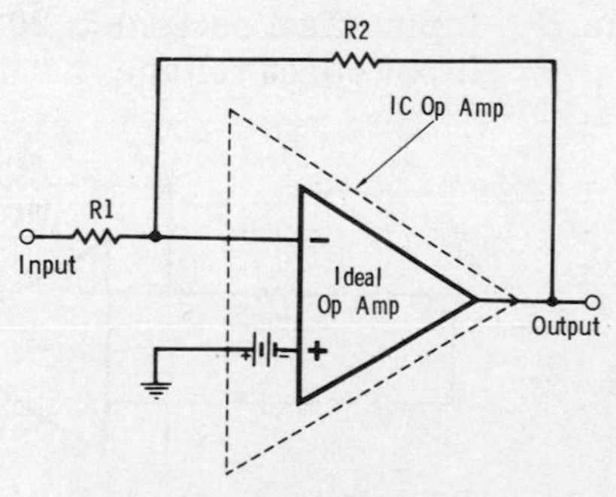

Fig. 4-4. Input offset voltage model for IC op amp.

The input offset voltage can be modeled as a voltage source in series with the +input lead of an ideal op amp, as seen in the inverting-amplifier circuit of Fig. 4-4. For this inverting-amplifier circuit, the output offset voltage due to the input offset voltage can be calculated as the response of a noninverting amplifier (Fig. 1-9A) to the input offset voltage, as shown in Fig. 4-5:

$$E_{os} = \frac{R_1 + R_2}{R_1} \times V_{os}$$

where,

E_{os} is the output offset voltage in volts,
V_{os} is the input offset voltage in volts,
R_1 and R_2 are in ohms.

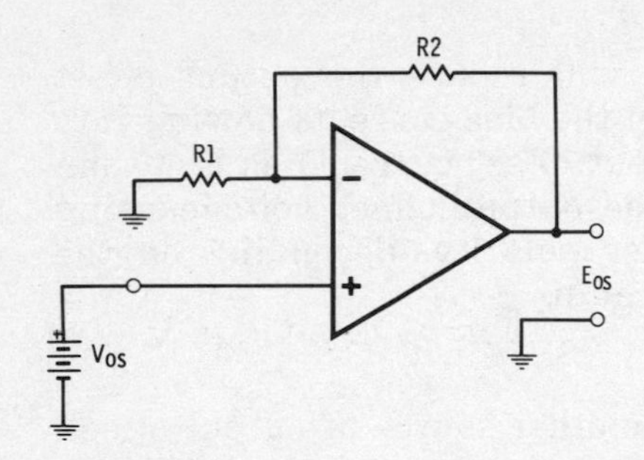

Fig. 4-5. Noninverting amplifier model for simple inverting or noninverting amplifier calculations.

Example Calculation of Output Offset Voltage

Fig. 4-6 shows an actual inverting-amplifier circuit using a 741 op amp. The output offset is due to two sources, as discussed in the previous sections.

Using the typical 741 data given in the Appendix:

Input bias current = 80 nA = 8×10^{-8} ampere.
Input offset current = 20 nA = 2×10^{-8} ampere.
Input offset voltage = 1 mV = 10^{-3} volt.

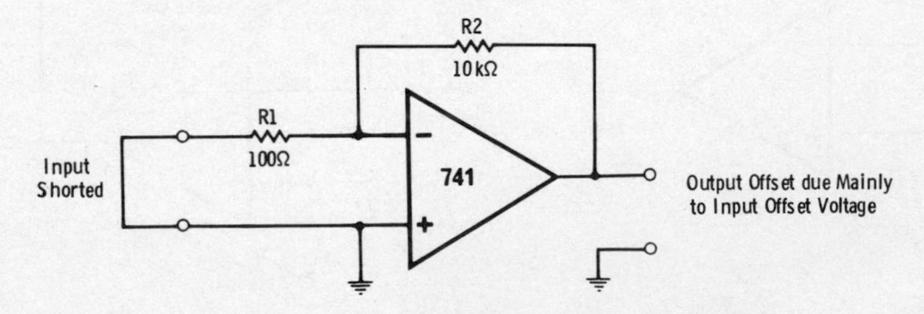

Fig. 4-6. Inverting amplifier with gain of 100.

The output offset due to input bias current is:

$$E_{os} = I_b \times R_2,$$
$$= 8 \times 10^{-8} \times 10^4,$$
$$= 8 \times 10^{-4} \text{ volt},$$
$$= 0.8 \text{ millivolt}.$$

The output offset due to input offset voltage is:

$$E_{os} = \frac{R_1 + R_2}{R_1} \times V_{os},$$
$$= \frac{100 + 10^4}{100} \times 10^{-3},$$
$$= 101 \times 10^{-3} \text{ volt},$$
$$= 101 \text{ millivolts}.$$

Since the two output offset voltages are caused by different sources, they may be of either polarity, so they may either add or subtract. In the worst case they will add, giving a total output offset of 101.8 millivolts, which for most applications is fairly small. The addition of resistor R_3, as shown in Fig. 4-2, will decrease the component of output offset due to bias currents to:

$$E_{os} = I_{os} \times R_2,$$
$$= 2 \times 10^{-8} \times 10^4,$$
$$= 2 \times 10^{-4} \text{ volt},$$
$$= 0.2 \text{ millivolt}.$$

For these component values this additional resistor does not seem worthwhile, but if resistors R_1 and R_2 were 5 MΩ and 10 MΩ, respectively, as shown in Fig. 4-7, the effect of resistor R_3 would become more important.

Recalculating using the different component values:

$$E_{os} = I_b \times R_2,$$
$$= 8 \times 10^{-8} \times 10^7,$$
$$= 0.8 \text{ volt}.$$

$$E_{os} = \frac{R_1 + R_2}{R_1} \times V_{os},$$
$$= \frac{5 \times 10^6 + 10^7}{5 \times 10^6} \times 10^{-3},$$
$$= 0.003 \text{ volt}.$$

$$E_{os} = I_{os} \times R_2,$$
$$= 2 \times 10^{-8} \times 10^7,$$
$$= 0.2 \text{ volt.}$$

The total worst-case output offset voltage will be decreased from 0.803 volt to 0.203 volt when resistor R_3 is used.

It is important to note that the output offset due to bias currents is dependent on the magnitude of the circuit resistances, whereas the output offset due to the input offset voltage is dependent on the dc closed-loop gain of the circuit. Thus, a low-gain amplifier circuit using large values of resistance (Fig. 4-7) will have an output offset dominated by the input bias current, whereas a high-gain amplifier circuit using small resistances (Fig. 4-6) will have an output offset dominated by the input offset voltage.

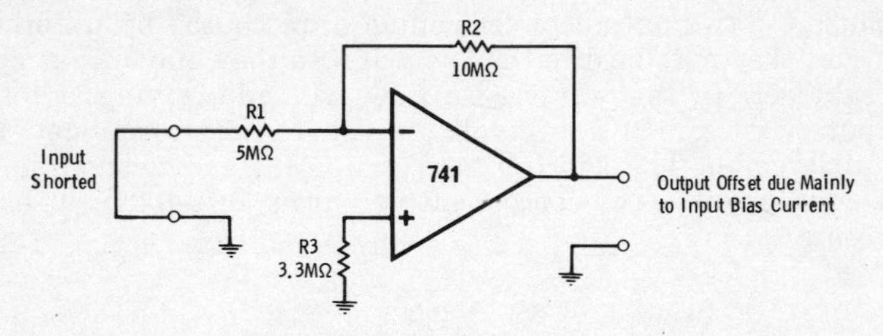

Fig. 4-7. Inverting amplifier with gain of 2.

Output Offset Calculations for Other Circuits

The techniques used to calculate the output offset of amplifier circuits are directly applicable to many other types of circuits. Since the output offset is strictly a dc phenomenon, it is dependent only on the resistances, and not the capacitances, in a circuit. Thus, the output offset for circuits with capacitors in parallel with resistors may be calculated by neglecting the capacitors. If no physical resistor is used in parallel with a capacitor, the leakage resistance of the capacitor should be used in the offset calculation.

As a practical example, consider the integrator circuit of Fig. 4-8. If R_2 were not placed in parallel with the feedback capacitor (C_1), the output offset due to the input offset voltage would be very high, since the leakage resistance of capacitors is normally very high. The purpose of R_2 is to reduce the output offset of the integrator by decreasing the dc closed-loop gain of the circuit.

Fig. 4-8. A practical integrator circuit.

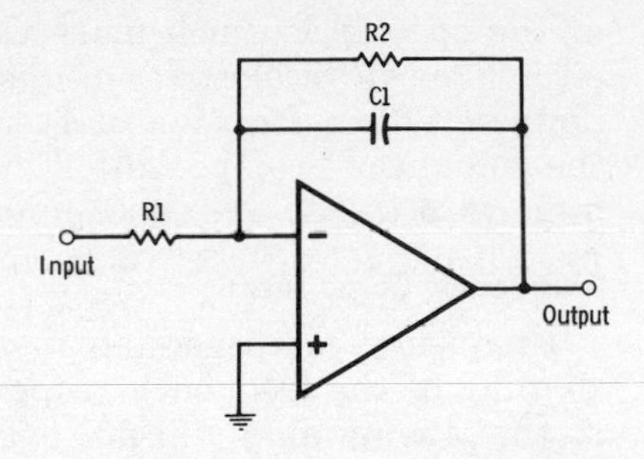

Output Offset Null Adjustment

Once a circuit using IC op amps has been designed with output offset in mind, the output offset should be reasonably small for most applications. However, if the output offset has to be *exactly* zero, a "zeroing" or "nulling" potentiometer may be used to zero the output offset. Although an external potentiometer may be used as shown in Fig. 4-9, many IC op amps have connections to internal points in the IC for output offset null adjustment, as shown in Fig. 4-10.

For an output offset with minimal drift, a nulling potentiometer should only be used after all other techniques have been applied to reduce the output offset. These techniques include using a resistor (R_3) in series with the +input, using an op amp with low input offset voltage and bias currents, using the smallest allowable values of circuit resistances, and using a design with minimum allowable closed-loop gain at dc.

FREQUENCY RESPONSE

In any circuit, the range of frequencies over which the circuit can be used is important. Circuits that use op amps are often designed with the assumption that the open-loop gain

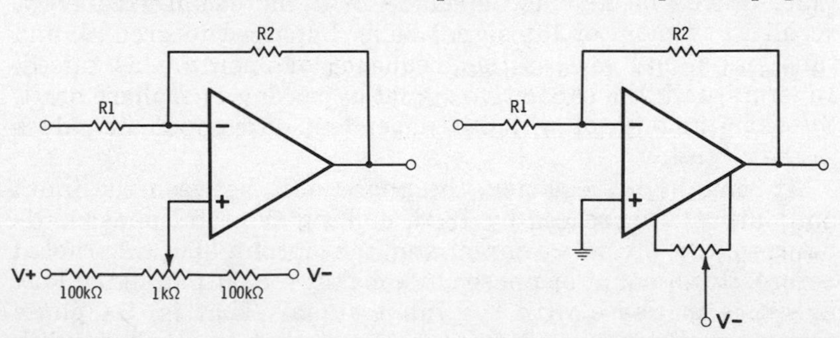

Fig. 4-9. Output offset null control for op amps without internal nulling provisions.

Fig. 4-10. Output offset null control for op amps with internal nulling provisions.

of the op amp is much more than the closed-loop gain of the circuit *for all frequencies of operation.* The gain of operational amplifiers as a function of frequency is largely dependent on the amount of "compensation" used to guarantee that a circuit built with the op amp does not oscillate.

Frequency Compensation

Frequency compensation in an operational amplifier is the shaping of the frequency response of the open-loop gain used, so that the op amp will not oscillate due to its internal phase shift when placed in a circuit. The op amp, when properly compensated, will act more ideally in that the op amp will not contribute to the instability of a circuit.

The need for frequency compensation is due to the phase shift in the op amp that is associated with the decrease in gain of the op amp at high frequencies. This phase shift, in effect, acts as a time delay in that the same point on a sine wave appears delayed in time, when phase shifted, by comparison to a reference signal that has not been phase shifted. It is possible to assign an angle to an amount of phase shift. Fig. 4-11 shows the effect of 90°, 180°, 270°, and 360° phase shift relative to a reference signal. Note that a 180° phase shift is just what an ordinary, ideal inverting amplifier has between its input and output signals, and that a 360° phase shift amounts to a delay of one full cycle, which makes it appear identical to the reference signal.

Fig. 4-12 shows the variation in the open-loop gain and phase shift of an op amp as a function of frequency. Note that at low frequencies, the phase shift between the input and output is 180°, as expected for an inverting amplifier. At higher frequencies, the phase shift increases while the gain decreases. Both effects are caused by circuit capacitances. These capacitances have a decreasing impedance with increasing frequency, resulting in more of the signal being bypassed to ground, and thus less signal gain as the frequency of operation is raised. Inherent with the capacitive signal bypassing is a phase shift, the exact amount of which is dependent on component values in the circuit.

At some high frequency, the phase shift between the input and output is increased by 180°, making the total phase shift between the inverting input and the output 360°. As noted before, this amount of phase shift makes the output signal look like it is in phase with the input signal. That is, its phase cannot be distinguished from a signal that is in phase with the input signal.

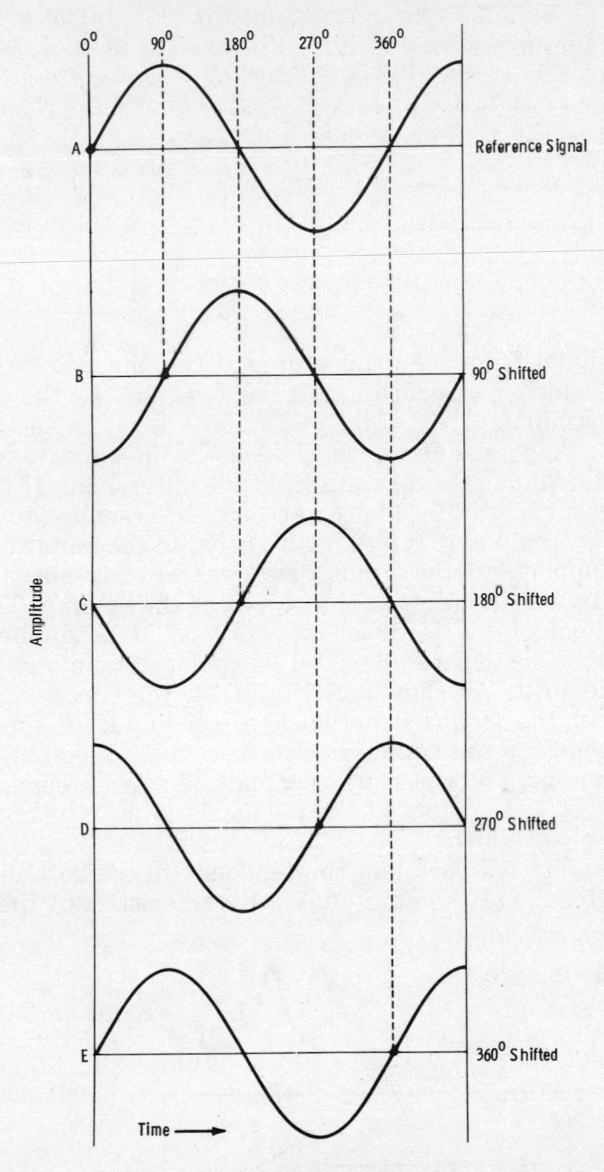

Fig. 4-11. Phase-shifted waveforms.

Fig. 4-13 shows an inverting amplifier with the circuit broken between the feedback network and the inverting input of the op amp, and with a sine-wave generator applied to the input of the op amp. The waveforms in this figure show the

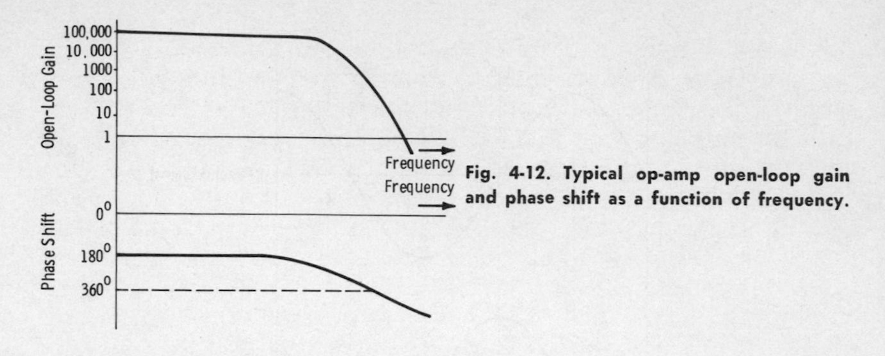

Fig. 4-12. Typical op-amp open-loop gain and phase shift as a function of frequency.

input signal from the generator and the phase-shifted output signal, which would ordinarily be fed back to the inverting input terminal. If the total phase shift through the amplifier is 360°, the signal fed back is identical in appearance to the generator signal, except for amplitude differences. If the signal coming out of the feedback network, in response to the generator at the op-amp input, is also equal to the amplitude of the signal applied to the input, the generator is not needed to maintain the sine wave at the output of the op amp. The signal coming out of the feedback network could be applied to the input, and the generator would be replaced by a signal source identical to it. As shown in Fig. 4-14, the connection of the output of the feedback network to the input of the op amp corresponds to the configuration of a typical op-amp circuit. Thus, an op-amp circuit will oscillate if a large enough signal is returned to the input at the frequency where the op amp has a 360° phase shift.

The signal fed back and the tendency to oscillate depend on two factors: (1) the gain-phase characteristics of the op amp

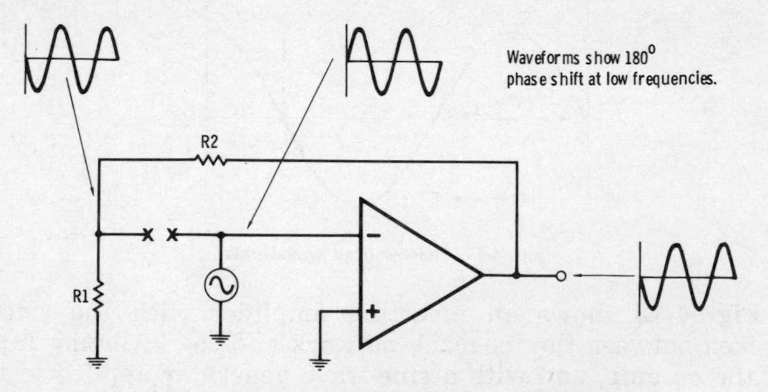

Fig. 4-13. Inverting amplifier with broken feedback path.

(Fig. 4-12) and (2) the feedback network. A low-gain inverting amplifier is more likely to oscillate than a high-gain one, because the low-gain amplifier has a larger percentage of the op-amp output signal fed back to the input. An integrator is usually very susceptible to oscillation because at high frequencies almost all the output is fed back to the op-amp input via the feedback capacitor, and it is at the high frequencies where the phase shift is 360°. It is not at all necessary to have a signal generator start the oscillation. If the gain is even slightly more than unity at the frequency where the total phase shift is 360°, the oscillation will grow out of any noise in the op amp. This oscillation will grow to an amplitude that is limited only by the power-supply voltages.

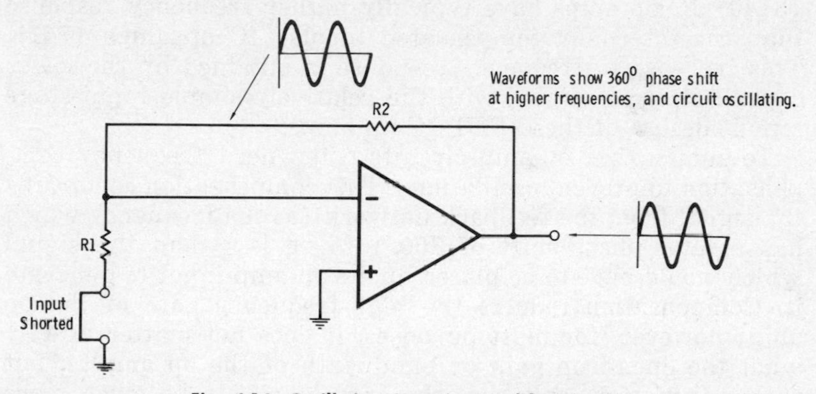

Fig. 4-14. Oscillating inverting-amplifier circuit.

It is certain that the amplifier will never oscillate, even if all the output is fed back to the input, if the gain of the op amp is less than unity at the frequency where the total phase shift is 360°. If the op amp has this gain-phase characteristic, it is called *unconditionally stable*. Any op amp can be made unconditionally stable by using some components to shape its frequency response in such a way that the gain is less than unity when the phase shift is 360°. This shaping is called *frequency compensation.*

Unfortunately, shaping does decrease the high-frequency response of the op-amp circuit. For highest frequency response, the minimum amount of frequency compensation necessary to achieve stability should be used. Since low-gain circuits have a larger percentage of the output fed back to the input than high-gain circuits do, it is possible that a low-gain circuit will have to be heavily compensated, whereas a high-gain circuit need not be compensated at all. (Heavy compensation usually implies large values of resistance and capacitance in the frequency-compensation network.)

Some IC op amps require external components to be added for frequency compensation, while others are internally compensated for unconditional stability. Usually the internally compensated op amps are more convenient to use, whereas the externally compensated op amps may be used at higher frequencies if a great deal of compensation is not required. Fig. 4-15 shows the frequency response that can be achieved in an inverting-amplifier circuit for several values of amplifier gain, using a minimally compensated 709 op amp. Note the decrease in amplifier bandwidth with the decrease in closed-loop gain.

The internally compensated BIFET (081) and BIMOS (3140) IC op amps have typically higher frequency response than the internally compensated bipolar IC op amps (741). This increase in frequency response is obtained by the lower capacitances associated with the relatively simple input-stage circuit design of these FET IC op amps.

To summarize, op-amp circuits often need frequency compensation to prevent oscillations. This compensation constrains the signal from the feedback network (at the frequency which has a total phase shift of 360°) to be less than the signal which would have to be placed at the op-amp input to generate it. Compensation reduces the high-frequency gain of the op amp; however, for most purposes, it does not matter exactly what the open-loop gain or bandwidth of the op amp is, but rather it matters only that the open-loop gain be much more than the closed-loop gain (at all operating frequencies) of the circuit in which it is used.

Slew Rate

If a large-signal square wave is put into the input of an op amp, there is a maximum speed at which the output voltage can

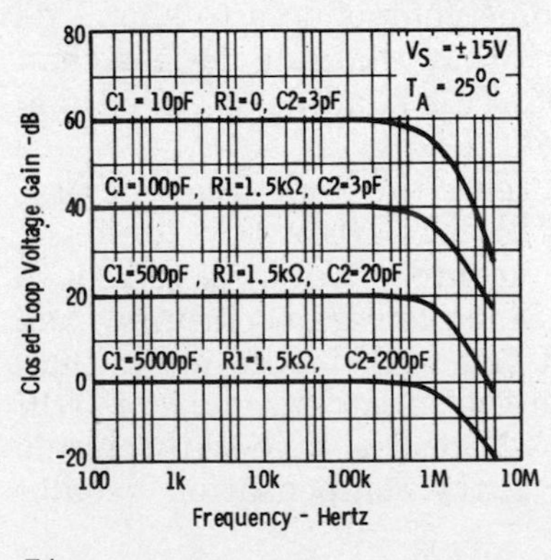

Fig. 4-15. Frequency response of type 709 op amp for various closed-loop gains and amounts of compensation.

swing from the most positive to the most negative, or vice versa. This maximum rate of change in the output voltage is called the *slew rate*. The slew rate can be used to compute an estimate of the time required to switch between the two maximum output levels:

$$\text{SWITCHING TIME} = \frac{\text{TOTAL POWER-SUPPLY VOLTAGE}}{\text{SLEW RATE}}$$

For a 741 powered by ±15-volt supplies:

$$\text{SWITCHING TIME} = \frac{30 \text{ volts}}{0.5 \text{ volt/microsecond}} = 60\mu\text{s}.$$

Increasing the frequency compensation decreases the slew rate, but slew rate is important mostly in switching circuits, such as comparators, in which frequency compensation is not required because no feedback is used.

The 741 is not as fast in slewing as some of the newer FET types of op amps. The 081, for example, is internally compensated and has a slew rate of 12 volts/microsecond, 24 times faster than the 741.

COMMON-MODE REJECTION

As discussed in Chapter 3, due to their differential input stage, op amps amplify mainly the difference between the signals applied to the two input terminals. Unfortunately, due to component mismatches and nonidealities in current sources, IC op amps amplify common-mode signals in addition to differential-mode signals. Typically, however, the gain for common-mode signals is much less than for differential-mode signals. The ratio of the differential-mode gain to the common-mode gain is the *common-mode rejection ratio* (CMRR). CMRR is often expressed in terms of decibels (dB):

$$\text{CMRR (ratio)} = \frac{\text{DIFFERENTIAL-MODE GAIN}}{\text{COMMON-MODE GAIN}}$$

$$\text{CMRR (dB)} = 20 \times \log \frac{\text{DIFFERENTIAL-MODE GAIN}}{\text{COMMON-MODE GAIN}}$$

Fig. 4-16 can be used to convert between CMRR (ratio) and CMRR (dB).

The common-mode input to the op amp shown in Fig. 4-17 gives an idea of the importance of the effect of CMRR. Using the appropriate 741 data (see Appendix):

$$\text{DIFFERENTIAL-MODE GAIN} = 200{,}000$$
$$\text{CMRR (dB)} = 90 \text{ dB}$$

Thus, using Fig. 4-16:

$$\text{CMRR (ratio)} = 31{,}000$$

$$\text{COMMON-MODE GAIN} = \frac{\text{DIFFERENTIAL-MODE GAIN}}{\text{CMRR (ratio)}},$$
$$= \frac{200{,}000}{31{,}000},$$
$$= 6.45.$$

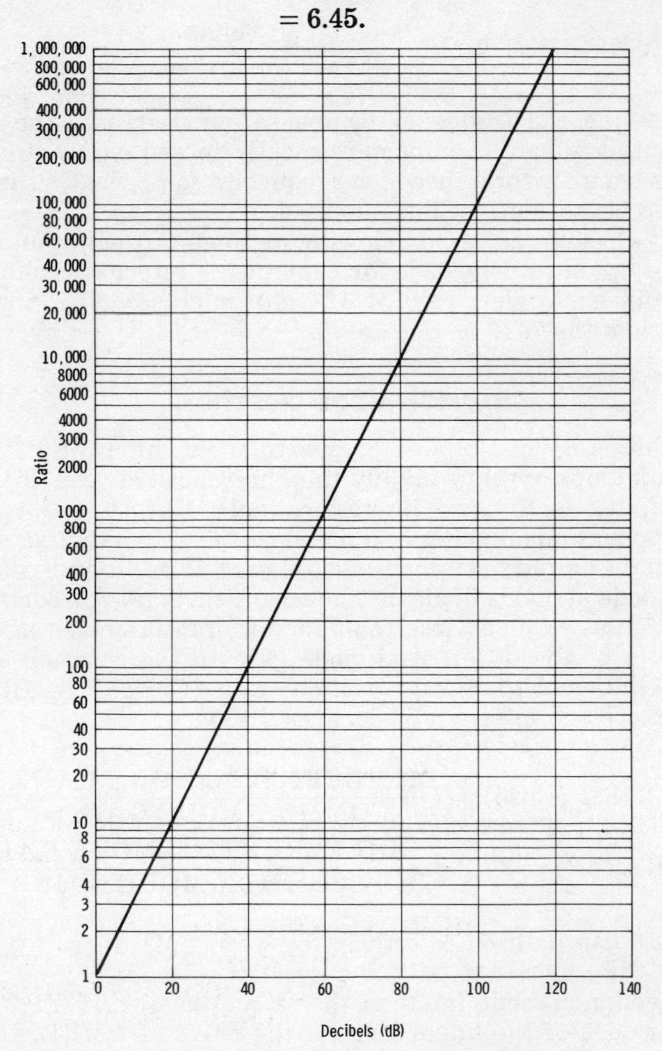

Fig. 4-16. CMRR ratio/dB conversion graph.

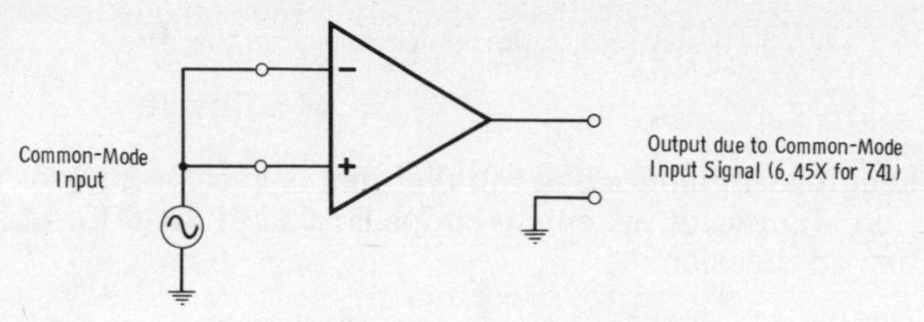

Fig. 4-17. Illustration of common-mode gain.

The output in response to the common-mode signal is thus 6.45 times the common-mode input voltage. In an amplifier circuit, the differential-mode gain acts in opposition to the common-mode output, because the common-mode output causes a differential signal to be fed back to the op-amp input via the feedback network. Thus, the ratio of the two types of gain (the common-mode rejection ratio) is used to determine the output in response to a common-mode input signal:

$$\text{CIRCUIT GAIN FOR COMMON-MODE SIGNAL} = \frac{\text{CIRCUIT GAIN FOR DIFFERENTIAL-MODE SIGNAL}}{\text{CMRR (ratio)}}$$

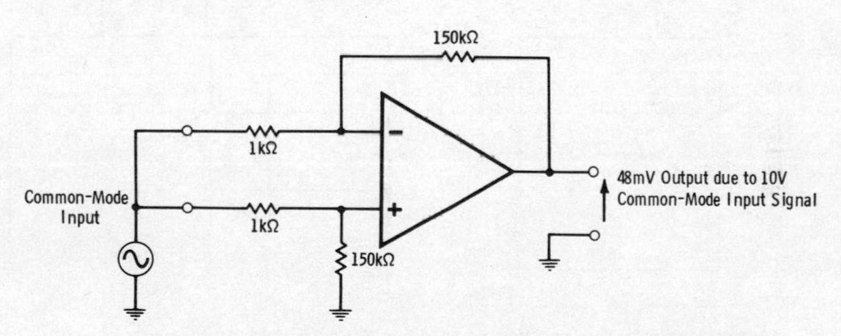

Fig. 4-18. Gain-of-150 differential amplifier circuit with output due to common-mode input signal.

The output in response to the common-mode signal applied to the differential-amplifier circuit (with a differential-mode gain of 150) shown in Fig. 4-18 can easily be found:

$$\text{CMRR (ratio)} = 31{,}000$$

$$\text{OUTPUT}_{\text{COMMON-MODE INPUT}} = \text{COMMON-MODE CIRCUIT GAIN} \times \text{COMMON-MODE INPUT SIGNAL}$$

For a 10-volt common-mode signal:

$$\text{OUTPUT}_{\text{COMMON-MODE INPUT}} = \frac{150}{31{,}000} \times 10,$$
$$= 48 \text{ millivolts.}$$

Although a common-mode input signal is greatly suppressed, it can be significant when op amps are used in differential-amplifier applications.

POWER SUPPLIES

Almost all op amps are designed to use two power supplies—one positive and one negative. Although some op amps operate on supplies of 5 volts or less, most are designed for +15-volt and −15-volt supplies (Fig. 4-19A). Usually the power-supply voltage is not very critical for op-amp performance, and ±9 volts, which is a convenient battery voltage, can be used to supply op-amp circuits. Typically, op-amp circuits can be easily battery-powered, because only a few milliamperes of current per op amp are required.

Sometimes a single power supply, either positive or negative, is all that is required to power op amps if some compromises can be made. Two schemes for this are shown in Figs 4-19B and 4-19C. The drawback to using one power supply is that

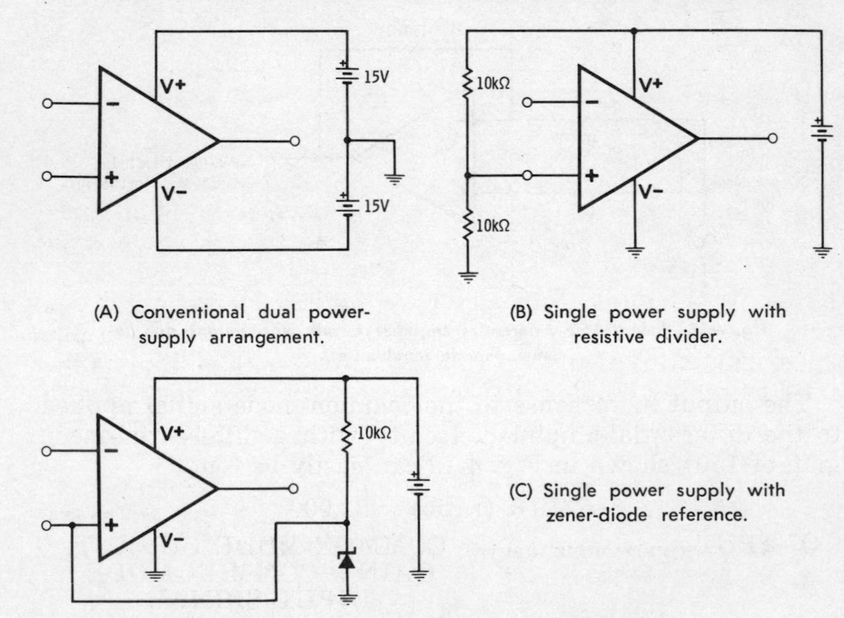

(A) Conventional dual power-supply arrangement.

(B) Single power supply with resistive divider.

(C) Single power supply with zener-diode reference.

Fig. 4-19. Typical power-supply arrangements.

interfacing with other circuits becomes difficult, because for "zero" input the output is half the power-supply voltage, and the "zero" input signal needs to be biased to half the supply voltage. However, these techniques may be used to avoid the need for two supplies, especially for ac-coupled circuits.

The BIMOS (3140) and CMOS (3130) IC op amps are unique in some power-supply configurations because the input-stage voltage level can go slightly below the negative power-supply voltage. For example, these op amps can be operated properly with the negative (V−) supply at ground (0 volts) and the positive (V+) supply at +10 volts even though the input varies from −0.2 volt to −0.3 volt, provided the output voltage is within the range of 0 to +10 volts.

NONLINEAR RESPONSES

Op amps may sometimes give an output signal far different from that which would be expected from the input signal. These outputs are generally undesirable responses, and it is the purpose of this section to identify them and show how to avoid them.

Latch-Up

Latch-up is the "sticking," or "locking up" of the output of an op amp when the maximum differential input voltage is exceeded. In the latched-up condition, the output is stuck at either the positive or negative maximum output voltage, and the input is ineffective in affecting the output. As discussed in Chapter 3, some of the newer op-amp designs, such as the 741, have circuit changes that virtually eliminate the latch-up problem by increasing the maximum input voltage of the op amp.

Current Clamping

Current clamping is the distortion in the op-amp output due to the current-limiting characteristics of the op-amp output stage. This distortion occurs in the newer op-amp designs because they have built-in current limiters in the output stage to protect the op amp from short circuits. Fig. 4-20 shows the

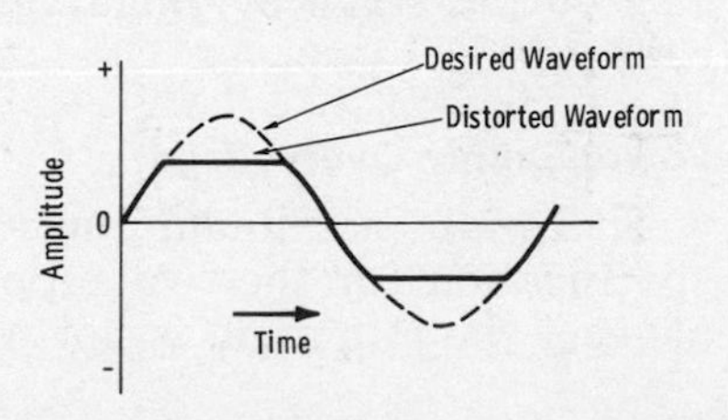

Fig. 4-20. Distortion of output due to current limiting.

effect of current clamping in the 741 op amp, which limits the output current to a maximum of 25 milliamperes.

The 3130 CMOS IC op amp has less output drive (only 3 to 10 milliamperes) due to the use of MOS transistors in the output amplifier stage. Also, MOS transistors have lower transconductance; therefore, they provide less current. These MOS transistors do provide output voltage swings that are much closer to the power supply than their bipolar transistor counterparts.

Voltage Clipping

Voltage clipping is the distortion in the op-amp output caused by the limiting of the maximum output voltage of the op amp. Typically, the maximum excursions of the output voltage are limited to approximately one volt less than the power-supply voltages. An input to an IC op amp, which would require an output excursion greater than the maximum output voltage (of either polarity) of the op amp, will result in clipping of the output voltage. Clipping of the output voltage has an effect on the output waveform similar to current clamping.

The 3130 IC op amp has the unique capability of providing output voltages to within a tenth of a volt of the power supplies. This feature can be useful in low supply-voltage applications.

Crossover Distortion

Crossover distortion is the ouput waveform distortion resulting from the class-B mode of bias used on a typical op-amp output amplifier stage. The 080 series BIFET op amp has a specially designed output stage to minimize crossover and harmonic distortion.

PRECAUTIONS WITH IC OP AMPS

As is true with any other device, IC op amps can be destroyed if they are not properly cared for. There are four common sources of destruction: (1) power-supply overvoltage, (2) input-stage overvoltage, (3) power-supply polarity reversal, (4) output-stage overloading, and (5) static-electricity discharges.

Power-Supply Overvoltage

There is a maximum and a minimum usable power-supply specification for the two supply voltages of every op amp. If at least the minimum is not provided, the op-amp circuit will just not function properly. If the maximum is exceeded, the collector-base junctions of the integrated-circuit transistors will break down, causing large currents to flow. Although the breakdown is not in itself necessarily destructive (as discussed in the section on diodes in Chapter 2), the large currents that flow as a result of the breakdown cause the destruction. The op-amp failure can take two forms: (1) overheating of the semiconductor chip, causing structural changes, or (2) destruction of the small bonding and connecting wires due to excessive current flowing through them.

Input-Stage Overvoltage

An excessive differential input voltage will break down the base-emitter junctions of the input stage of op amps. In simpler input-stage circuits, this can occur at 5 volts. If large currents flow due to this breakdown, the op amp will be destroyed through effects similar to power-supply overvoltage. This type of failure can cause a seemingly mysterious destruction of an op amp in an integrator-type circuit that has a large-value capacitor in the feedback loop. During power-supply transients, such as turning the circuit on or off, the feedback capacitor can destroy the input stage with its stored charge if the maximum differential input voltage is less than the power-supply voltage. The 709, for example, is vulnerable to this type of failure.

Power-Supply Reversal

Due to the form of isolation used to maintain separation between components (reverse-biased junctions) in an integrated circuit, op amps will usually be destroyed if the power-supply polarity is reversed. The destruction is due to large currents that flow through the isolation junctions, which become forward-biased if the supplies are reversed. This type of failure is most frequent in battery-operated equipment in which the batteries have been erroneously inserted in the

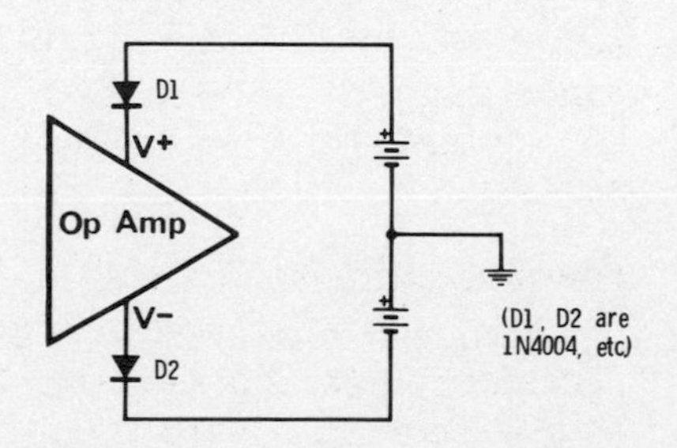

Fig. 4-21. Diodes D_1 and D_2 prevent op-amp failure if power-supply polarity is reversed.

wrong direction. Two diodes, as shown in Fig. 4-21, can be used to prevent this type of failure.

Output-Stage Overloading

Due to inefficiencies, the output stage of an op amp generates heat when delivering power to a load. Most IC op amps can safely dissipate about ½-watt of power at room temperature. The newer-designed op amps have output-current clamping. Current clamping limits the maximum output current (hence the power dissipation in the op amp) to a safe level, even if the output is accidentally short-circuited. Op amps that do not have current clamping, such as the 709, should be used with a resistor in series with the output to prevent accidental excessive power dissipation in the output stage.

Static-Electricity Discharges

Static electricity, created by rubbing electrons off of an insulator, can create thousands of volts when placed on the few picofarads of input capacitance of FET IC op amps. JFETs are more durable when exposed to this form of abuse than MOSFETs are, because there is a thin insulating oxide to rupture in the MOSFETs that is not in the JFETs. Thus, one should use particular caution when using 3130 and 3140 op amps to ensure that no static discharges are presented to the device input.

CHAPTER 5

Linear Applications

An op amp is operating in its linear range when the output of the op amp is directly proportional to the input. For the most common IC op amps, used with a ±15-volt supply, the op amp is in its linear range whenever its output is between −14 and +14 volts. When the output reaches its maximum or minimum excursion, however, the op amp is said to be saturated and is operating in its nonlinear range.

This chapter presents circuit applications in which IC op amps are used in their linear range. Since the op amp is a very high gain device, nearly all linear applications require the use of negative feedback.

GENERAL DESIGN CONSIDERATIONS

The use of the ideal op amp in linear applications was discussed in Chapter 1. However, as pointed out in Chapter 4, there are several considerations that apply to IC op-amp design that do not apply to design with the ideal op amp. The three most important considerations are output offset, frequency compensation, and gain.

Output Offset

Input bias current, input offset current, and input offset voltage are the three factors that can cause an output offset voltage. If the output offset is small, it is seldom a problem; the output can be capacitively coupled to the next stage, or a potentiometer can be added to the circuit to null the output offset. If the output offset is large, however, the input signal may cause the amplifier to swing into saturation, resulting in distortion of the output signal. In an extreme case, the output offset itself may drive the op amp into saturation.

Large output offset voltages can be avoided by proper circuit design. The effect of input bias current can be minimized by adding a single resistor from the +input of the op amp to ground. The effect of input offset current can be reduced by deriving the feedback signal from a voltage divider across the output. The input offset voltage is usually only troublesome when building a very high gain dc amplifier. These problems will be discussed in more detail in the descriptions of the sample circuits in this chapter.

Frequency Compensation

Due to the phase shift in IC op amps, negative feedback can cause undesired oscillation. The oscillation can be stopped by compensating the op amp. This is done by adding one or more frequency-compensation components to the compensation terminals of the op amp. The values of the compensation components depend on the closed-loop gain of the circuit, and these values are given in the op-amp specification sheets (see Appendix). Some op amps, like the 741, are internally compensated and never require additional compensation components. It is important to remember that frequency compensation decreases the high-frequency open-loop gain of the op amp.

Gain

The gain of IC op amps decreases as the operating frequency increases. As a general "rule-of-thumb," when designing a negative-feedback circuit the typical open-loop gain of the op amp should be at least 50 times greater than the closed-loop gain for all frequencies at which the circuit is used.

The 741 op amp, for example, has a typical gain of 200,000 at frequencies below 10 Hz, but a gain of only 100 at 10 kHz. The 741 is not a good op amp to use if high gain is required at high frequencies. The reason for this is that the 741 is internally compensated to be unconditionally stable. When the 741 is used in a circuit with high closed-loop gain, it is compensated more than is necessary for stable operation; this results in a high-frequency performance that is compromised more than is necessary. When high gain is needed at high frequencies, op amps using external compensation, such as the 101 or the 748, are a better choice than internally compensated op amps.

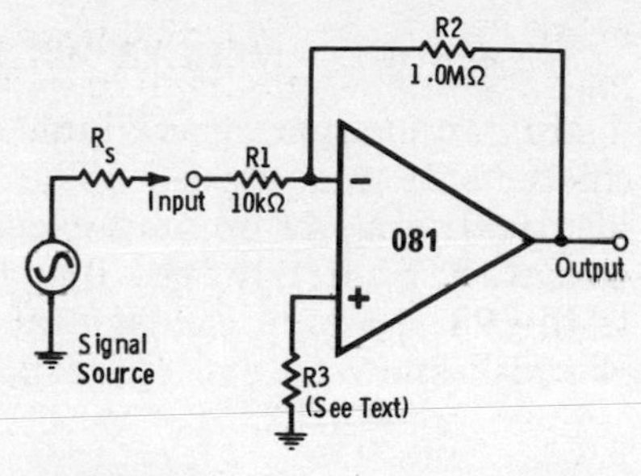

Fig. 5-1. Inverting amplifier with gain of 100.

INVERTING AMPLIFIER

Fig. 5-1 shows an inverting-amplifier circuit with an input impedance of 10 kΩ and a gain of 100. This circuit is very much like the circuit for the ideal op amp discussed in Chapter 1, with the addition of a resistor (R_3) between the +input and ground. The purpose of R_3 is to minimize the output offset due to the input bias current. The value of R_3 depends on the output resistance (R_s) of the input-signal source.

Input bias current drawn by the −input of the op amp flows through R_2 and the series combination of R_s and R_1. The bias current causes a voltage drop across these resistors with a resultant voltage appearing at the −input of the op amp. Resistor R_3 is chosen so that the voltage drop across it due to bias current drawn by the + input is exactly equal to the voltage drop at the −input. These two voltages cancel each other, producing no net output offset. Assuming that the bias current drawn by each input are about the same, R_3 can be calculated from the formula:

$$R_3 = \frac{(R_1 + R_s) \times R_2}{R_1 + R_s + R_2}$$

If the source output resistance is very low (for example, if the source is the output of another op amp), the formula for R_3 becomes:

$$R_3 = \frac{R_1 \times R_2}{R_1 + R_2}$$

If the source is capacitively coupled to the amplifier circuit, then no bias current can flow through R_s or R_1, and the formula for R_3 simplifies to:

$$R_3 = R_2$$

No frequency-compensation components are shown in Fig. 5-1 because the 081 op amp is internally compensated. Better high-frequency performance can be achieved by using a 080 op amp with external compensation.

INVERTING AMPLIFIER WITH EXTERNAL COMPENSATION

Many op amps are available either with or without a built-in compensation capacitor. The 748 op amp, for example, is identical to a 741 op amp except that it is not internally compensated. Similarly, the 080 FET-input op amp is identical to the 081 op amp except that it lacks internal compensation. Fig. 5-2 shows a 080 op amp with an external compensation

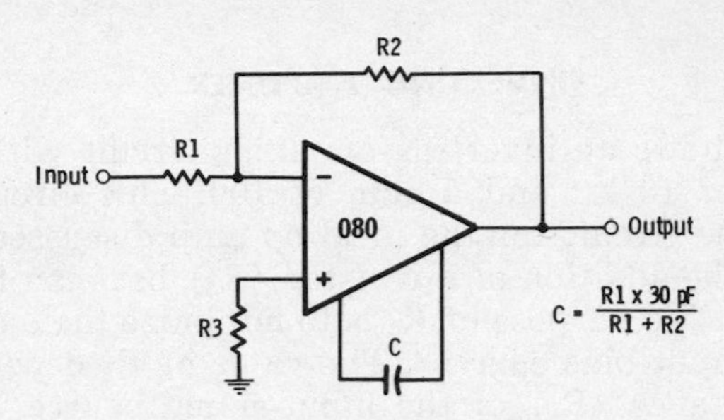

Fig. 5-2. Inverting amplifier with external compensation.

capacitor. The value of the compensation capacitor is given by the formula:

$$C = \frac{R_1 \times 30\text{pF}}{R_1 + R_2}$$

As the closed-loop gain of the circuit increases, smaller compensation capacitors can be used, which provide improved high-frequency response compared to circuits using internally compensated op amps.

INVERTING AMPLIFIER WITH OFFSET NULL ADJUSTMENT

When using operational amplifiers, it is often desirable to adjust the output offset voltage to a very low value. Most op amps have the provision for adding an offset null control so that this adjustment can be made. The offset null adjustment circuit differs from one op amp to another and is normally shown in the manufacturer's specification sheet. The offset null adjustment circuit for a 080 op amp is shown in Fig. 5-3.

NONINVERTING AMPLIFIER

The circuit diagram of a noninverting amplifier is shown in Fig. 5-4. This circuit is very similar to that of the inverting amplifier, except that the input signal is applied to the non-

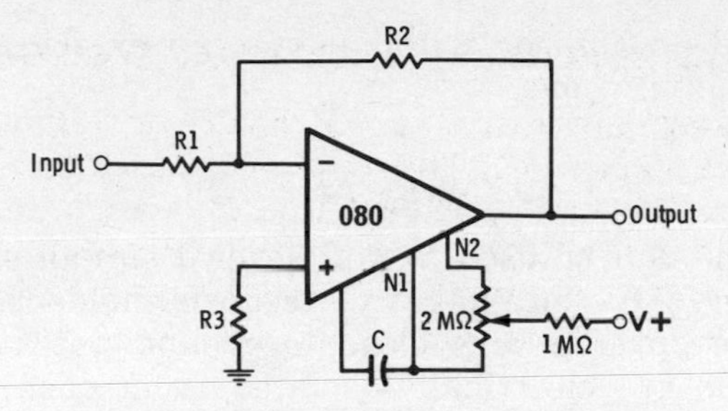

Fig. 5-3. Inverting amplifier with offset null adjustment.

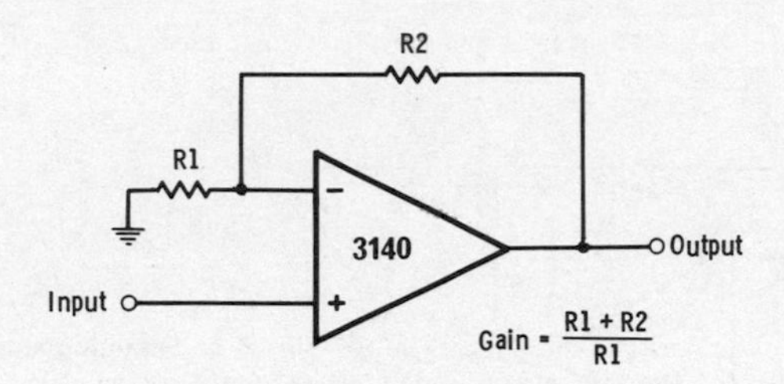

Fig. 5-4. Noninverting amplifier.

inverting input of the op amp. The gain of this noninverting circuit is given by the formula:

$$\text{GAIN} = \frac{R_1 + R_2}{R_1}$$

VOLTAGE FOLLOWER

The voltage-follower circuit, shown in Fig. 5-5, is a type of noninverting amplifier. The feedback resistance is zero (since the output is tied directly to the inverting input) so the gain of the circuit is unity. The voltage-follower circuit has a high input impedance, a low output impedance, and is a good choice whenever a buffer amplifier is needed. The voltage-follower circuit should be designed either with an internally

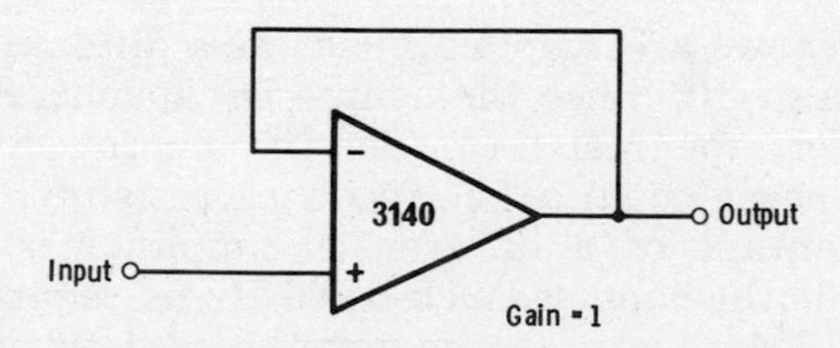

Fig. 5-5. Voltage-follower circuit.

compensated op amp or with an op amp externally compensated for a gain of one.

SUMMING AMPLIFIER

An op amp can be used to add several signals together, as shown in Fig. 5-6. Since negative feedback holds the inverting input of the op amp very close to ground potential, all the input signals are electrically isolated from each other. The output of the amplifier is just the inverted sum of the input signals.

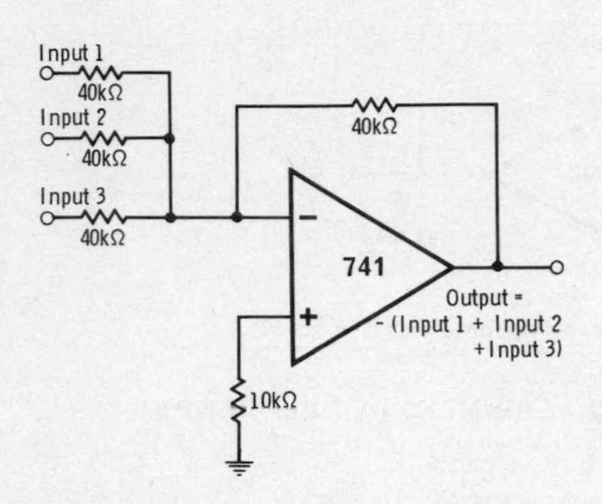

Fig. 5-6. Summing-amplifier circuit.

The resistor from the noninverting input to ground is chosen as the parallel combination of the input and feedback resistances, assuming low source resistances. The circuit can have gain if the size of the feedback resistance is increased, and the signals can be added with different gains if different input resistors are used for the different input signals.

One application for the summing amplifier is in the mixing of audio signals. Potentiometers are used for the input resistors to adjust the relative volume of each of the input channels.

DIFFERENTIAL AMPLIFIER

Fig. 5-7 shows a 741 op amp used as a differential amplifier with a gain of 10. Since the source is capacitively coupled to the amplifier, the resistor from the noninverting input to ground is chosen equal to the feedback resistor.

The advantage of a differential amplifier is that signals common to both channels, such as 60-Hz noise, are suppressed. Differential amplifiers are commonly used when very small voltages, such as bioelectric potentials, need to be measured.

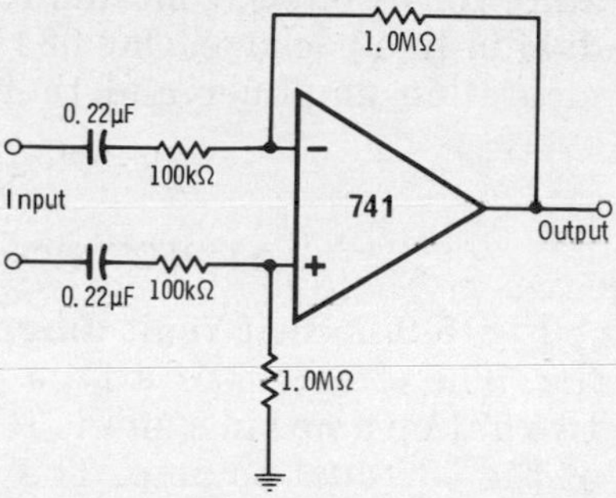

Fig. 5-7. Differential amplifier with gain of 10.

INSTRUMENTATION AMPLIFIER

The instrumentation amplifier shown in Fig. 5-8 is one of the most common designs using operational amplifiers. The instrumentation amplifier has excellent common-mode rejection and high input impedance. A key feature of this amplifier

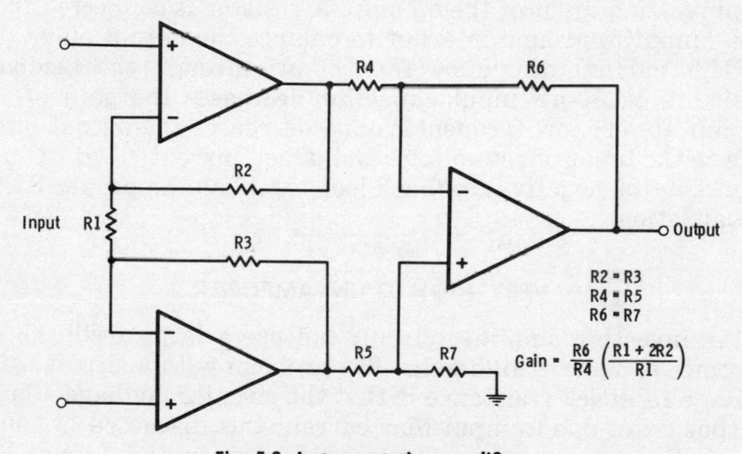

Fig. 5-8. Instrumentation amplifier.

is that the common-mode rejection of the first stage is not adversely affected by poor matching of the feedback resistors, R_2 and R_3. The gain of this instrumentation amplifier is given by the formula:

$$\text{GAIN} = \frac{R_6}{R_4}\left(\frac{R_1 + 2R_2}{R_1}\right)$$

When building circuits that use several op amps, it is often convenient to make use of integrated circuits that contain two or more op amps in a single package. The type 084 quad op amp, for example, contains four type 081 op amps in a single, dual in-line package. One 084 could be used to build the instrumentation amplifier and there would still be an op amp left over.

STEREO PHONO PREAMPLIFIER

Fig. 5-9 is the circuit diagram for a stereo phono preamplifier. The circuit uses a type 072 dual op amp which contains two 071 op amps in a single IC. The 072 is the low-noise version of the 082 dual op amp. The 072 was designed specifically for audio applications where low noise is important. The circuit would work equally well using two 071 op amps in place of the 072. The preamp was designed to be used with a magnetic cartridge and is RIAA equalized.

The circuit of Fig. 5-9 is basically a noninverting amplifier. A 47-kΩ resistor is used to load the magnetic cartridge, and the voltage across this resistor is capacitively coupled to the noninverting input of the op amp. A resistor is connected from the +input to ground in order to balance the output offset due to bias current drawn by the −input through the feedback resistors. A 50-μF input capacitor decreases the gain of the circuit at very low frequencies and decreases the output offset due to the input offset voltage and offset current. Two RC networks in the negative feedback loop establish the proper RIAA equalization.

VERY HIGH GAIN AMPLIFIER

An inverting amplifier circuit can use a large feedback resistance to achieve high gain. The problem with a circuit using a large feedback resistance is that the circuit may have a large output offset due to input bias currents, as discussed in Chapter 4. Even when a resistor is used between the +input and ground, the input offset current may still cause a large output offset. The amplifier circuit shown in Fig. 5-10 avoids this problem by deriving its feedback signal from a voltage divider (R_4, R_5) across the op-amp output, rather than directly from the output. The gain of this circuit is given by:

$$\text{GAIN} = \frac{R_2 + \left(\dfrac{R_2 \times R_5}{R_4}\right) + R_5}{R_1} = 5150$$

If a conventional circuit were used, such as that of Fig. 5-1, a 51.5-MΩ feedback resistor would have been required to

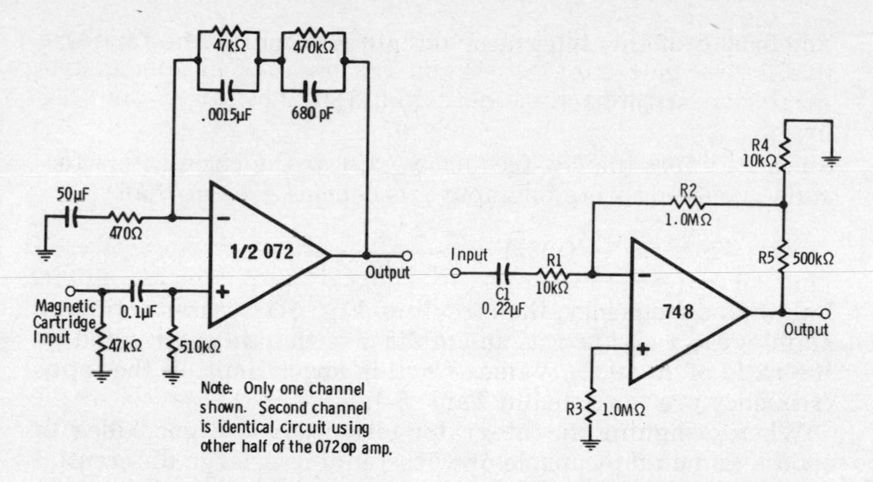

Fig. 5-9. Stereo phono preamplifier circuit.

Fig. 5-10. Very high gain inverting-amplifier circuit.

achieve this high gain. Notice that the input to the circuit is capacitively coupled (through C_1) in order to minimize the output offset due to input offset voltage.

INTEGRATOR

The integrator circuit shown in Fig. 5-11 has two components not used in the ideal integrator of Chapter 1. The purpose of R_3, which is equal in value to the parallel combination resistance of R_1 and R_2, has been discussed earlier in this chapter. The purpose of R_2 is to limit the low-frequency gain of the circuit. If the low-frequency gain were not limited, the small dc offset voltage would be integrated, and the integrator would ramp up (or down) until the op amp was saturated.

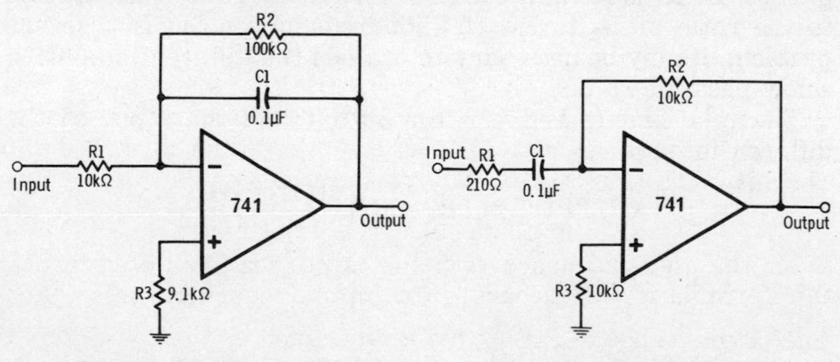

Fig. 5-11. Integrator circuit.

Fig. 5-12. Differentiator circuit.

The output of this integrator circuit is given by the formula:

$$\text{OUTPUT} = -\frac{1}{R_1 \times C_1}\int (\text{INPUT})\, dt$$

Since R_2 limits the low-frequency gain of the circuit, this formula is only accurate for input frequencies greater than:

$$f(\text{Hz}) = \frac{1}{2\pi \times R_2 \times C_1}$$

Below this frequency, the circuit of Fig. 5-11 ceases acting as an integrator and acts as an amplifier with gain determined by the ratio of R_2 to R_1. Values for this lower limit on the input frequency are tabulated in Table 5-1.

When designing an integrator circuit, it is a good idea to choose an op amp capable of withstanding a large differential input voltage, since voltage transients can be coupled to the input through the feedback capacitor. The design of the input stages of the 741 op amp, as discussed in Chapter 4, makes it a good choice for use as an integrator.

DIFFERENTIATOR

In the differentiator circuit of Fig. 5-12, resistors R_1 and R_3 have been added to the ideal differentiator circuit of Chapter 1. Resistor R_3 is set equal to R_2 to balance output offset due to input bias current. Resistor R_1 is used to limit the high-frequency gain of the circuit.

Differentiator circuits are often avoided in instrumentation systems because high-frequency noise, even though it may be low in amplitude, can have a very large derivative. Recall from Chapter 1 that differentiators respond to *changes* in the input signal, and high-frequency noise changes very rapidly. The purpose of R_1 is to limit the high-frequency gain of the circuit to the ratio of R_2 to R_1. If high-frequency noise is a severe problem, it may be necessary to precede the differentiator with a low-pass filter.

The relationship between the output and the input of the differentiator is:

$$\text{OUTPUT} = (R_2 \times C_1)\,\frac{d(\text{INPUT})}{d(\text{TIME})}$$

Since the high-frequency response is purposely limited by R_1, this formula is only accurate for input frequencies less than:

$$f(\text{Hz}) = \frac{1}{2\pi \times R_1 \times C_1}$$

Some values for this upper limit on the input frequency are given in Table 5-1.

PRECISION HALF-WAVE RECTIFIER

Since silicon diodes must be forward-biased to about 0.7 volt and germanium diodes to about 0.3 volt before they will conduct effectively, semiconductor diodes alone are not good rectifiers for small signals. When silicon diodes are used in the feedback loop of an op amp, however, a circuit can be built that will precisely rectify even very small input signals. Such a circuit is shown in Fig. 5-13A.

Notice that the output of this circuit is not taken from the output of the op amp, but rather from the connection between R_2 and D_2. The operation of this precision half-wave rectifier can be understood in terms of the equivalent circuit of Fig. 5-13B. When the input signal is positive, all the feedback current will flow through D_1, and the output voltage (of the circuit, not of the op amp) will be zero. But when the input signal is negative, all the feedback current will flow through D_2 and R_2, and an output voltage (the voltage drop across R_2) will appear at the output of the circuit. Because of the high gain of the op amp, even a very small negative input signal is adequate to forward bias D_2, so the circuit will precisely rectify even very small input signals.

PRECISION FULL-WAVE RECTIFIER

Fig. 5-14 is the circuit diagram of a precision full-wave rectifier. This circuit consists of a precision half-wave rectifier (Fig. 5-13A) followed by a summing amplifier (Fig. 5-6).

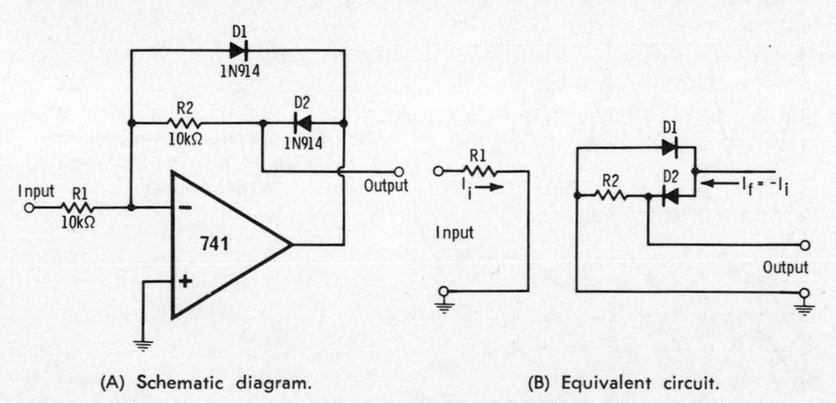

(A) Schematic diagram. (B) Equivalent circuit.

Fig. 5-13. Precision half-wave rectifier.

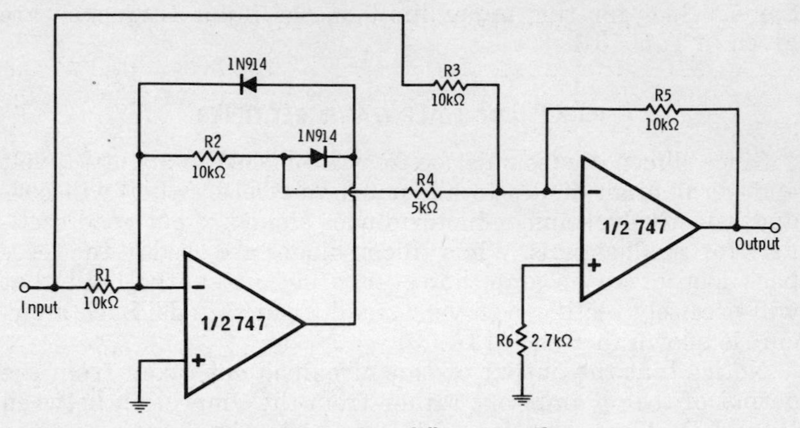

Fig. 5-14. Precision full-wave rectifier.

The principle of operation of the precision full-wave rectifier circuit can be seen from the waveforms shown in Fig. 5-15. Since the input resistors of the summing amplifier are selected with R_3 twice the value of R_4, the original input signal (Fig. 5-15A) is added to twice the output of the precision half-wave rectifier (Fig. 5-15B). The sum of these two waveforms is shown in Fig. 5-15C. Because the summing amplifier is also an inverting amplifier, the output waveform of the precision full-wave rectifier is the inverse of the waveform of Fig. 5-15C and is shown in Fig. 5-15D.

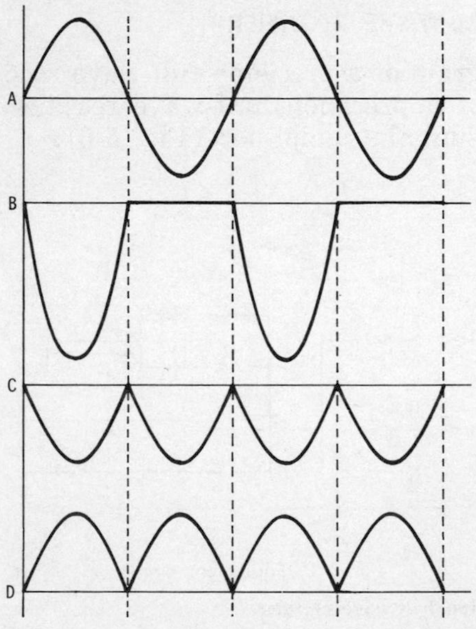

Fig. 5-15. Waveforms demonstrating operation of precision full-wave rectifier circuit.

VOLTAGE-REGULATED POWER SUPPLY

Fig. 5-16 is the circuit diagram of a voltage-regulated power supply. The circuit uses a type 741 op amp as a noninverting amplifier.

The input voltage to the noninverting amplifier is held constant by a zener diode, and the gain of the amplifier is given by the formula derived in Chapter 1:

$$\text{GAIN} = \frac{R_1 + R_2}{R_1}$$

The output voltage is varied by R_2. When a 3.6-volt zener diode is used for D_1, and a 470-ohm resistor is used for R_1, the output voltage is given by the formula:

$$\text{OUTPUT VOLTAGE} = 3.6 + (7.7 \times R_2)$$

where,

R_2 is in units of kilohms.

This output voltage will remain very stable, even with large changes in the output load resistance. Since the output of the 741 is current limited, a power transistor (Q_1) is used to boost the allowable output current to 0.5 ampere.

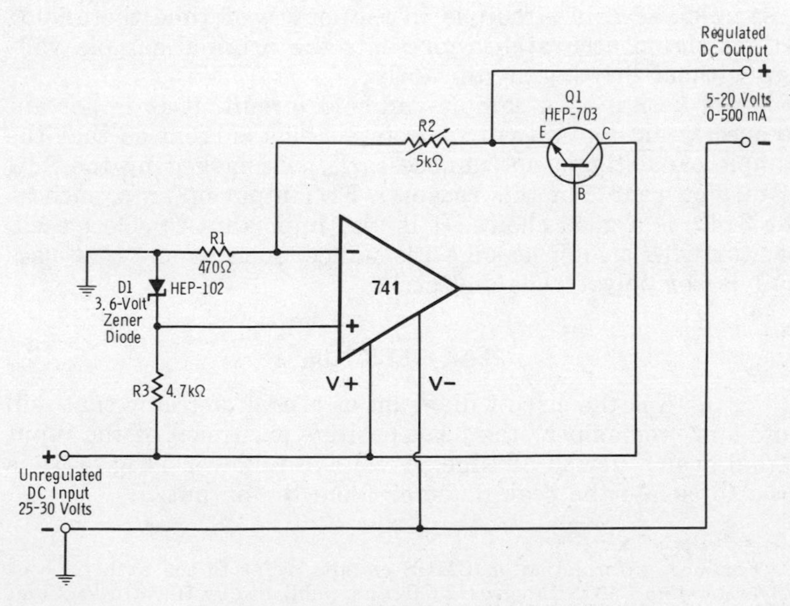

Fig. 5-16. Voltage-regulated power supply.

SAMPLE-AND-HOLD AMPLIFIER

A sample-and-hold amplifier is a circuit that makes it possible to sample the value of a time-varying analog signal at a particular point in time, and to retain that value of the signal. The sample-and-hold circuit is shown in Fig. 5-17. When a control signal is applied to the 4066 CMOS analog switch*, the sample capacitor charges up to the value of the analog input signal. When the control signal is removed, the analog input

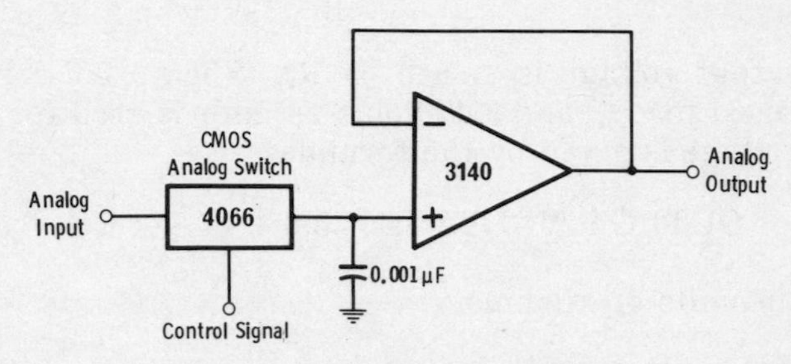

Fig. 5-17. Sample-and-hold amplifier.

is effectively disconnected from the sample capacitor. A 3140 op amp connected as a voltage follower is used to isolate the capacitor from the analog output. The measure of quality of a sample-and-hold circuit is in the length of time the analog output signal accurately represents the original sample voltage without drifting significantly.

When designing a sample-and-hold circuit, it is important to specify an op amp with low input bias current so that the sample capacitor is not unnecessarily discharged by the flow of bias current. For this reason a FET-input op amp, such as the 3140, is a good choice. It is also important to select a capacitor with a high leakage resistance, such as one that uses a mylar or polystyrene dielectric.

PEAK DETECTOR

Fig. 5-18 is the circuit diagram of a peak detector that will note and "remember" the peak positive excursion of the input signal over a period of time. A voltage follower (Fig. 5-5) is used to isolate the peak detector from the output.

---

* For more information on CMOS circuits, refer to the authors' book *Understanding CMOS Integrated Circuits*, published by Howard W. Sams & Co., Inc., 1975.

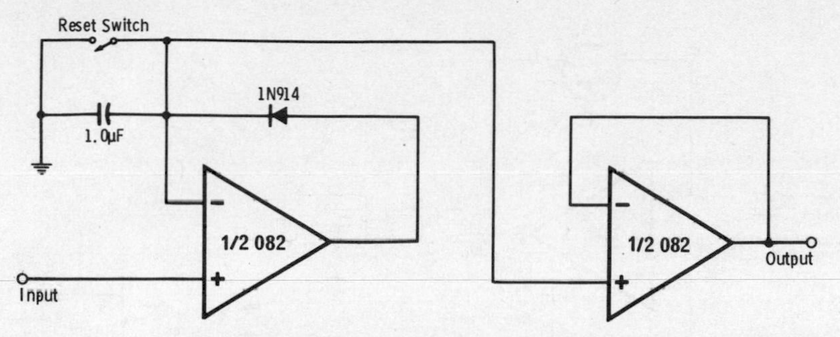

Fig. 5-18. Peak detector.

When a positive voltage is applied to the input of a recently reset peak detector, the noninverted output of the op amp forward biases the feedback diode. Even a very small voltage will forward bias the diode, in spite of the diode voltage drop, due to the high gain of the op amp. The capacitor will charge through the forward-biased diode until the differential inputs of the op amp are nearly at the same potential. Whenever the voltage at the +input exceeds the voltage at the −input, the capacitor will charge to this new peak value. Whenever the input voltage goes below the capacitor voltage, the op amp will swing into negative saturation, and the feedback diode will be reverse biased. The voltage across the capacitor is always equal to the greatest positive voltage that has been applied to the input since the circuit was last reset.

The memory time of the peak detector is typically several minutes; however, the capacitor will slowly discharge due to leakage currents in the capacitor, diode leakage current, bias current required by the op amps, and the slight loading effect of the voltage follower.

LOGARITHMIC AMPLIFIER

Fig. 5-19 is the circuit diagram of a *logarithmic* amplifier. The output of a logarithmic amplifier is proportional to the logarithm of the input. Logarithmic amplifiers are useful for converting amplitudes to decibels, for making compressor circuitry, and for building analog multipliers (Fig. 5-20).

The voltage across the feedback transistor in the logarithmic amplifier depends logarithmically on the feedback current. A power transistor is used to reduce the series feedback resistance. As discussed in Chapter 1, the magnitude of the current flowing through the feedback component is equal to the magnitude of the current flowing through the input component.

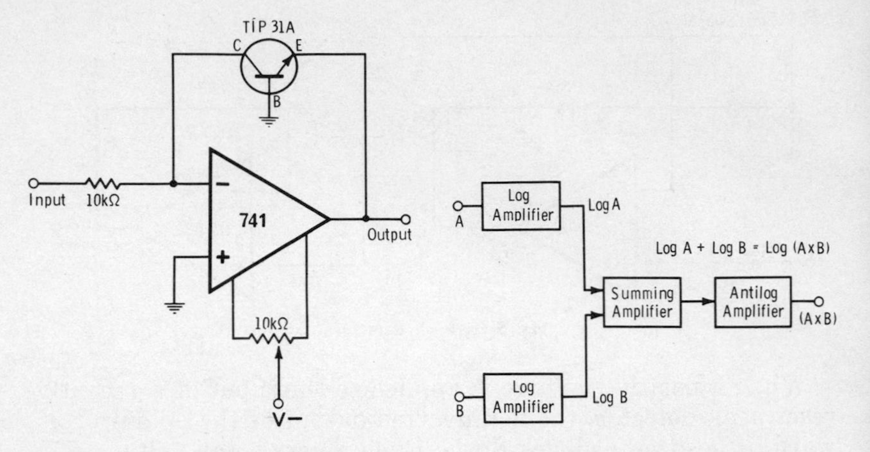

Fig. 5-19. Logarithmic amplifier. **Fig. 5-20. Analog multiplier.**

In the logarithmic amplifier, the input component is a resistor, and the feedback component is a grounded-base transistor. Since the input current is proportional to the input voltage, the feedback current is also proportional to the input voltage, and the output voltage is proportional to the logarithm of the input voltage.

The logarithm of one is zero. The op-amp offset null is used to zero the output of the op amp for one unit of positive signal at the input (the size of the unit used corresponds to the zero decibel reference).

An amplifier that has an input voltage proportional to the logarithm of the output voltage is called an *antilogarithmic* amplifier. An antilogarithmic amplifier can be made by using a grounded-base transistor as the input component and a resistor as the feedback component. An antilogarithmic amplifier is shown used in the block diagram of an analog multiplier (Fig. 5-20).

SINE-WAVE OSCILLATOR

Fig. 5-21 is the circuit diagram of a sine-wave oscillator using a single operational amplifier. The frequency of oscillation is determined by the formula:

$$f\,(\text{Hz}) = \frac{1}{2\pi RC}$$

Some standard values for this formula are tabulated in Table 5-1.

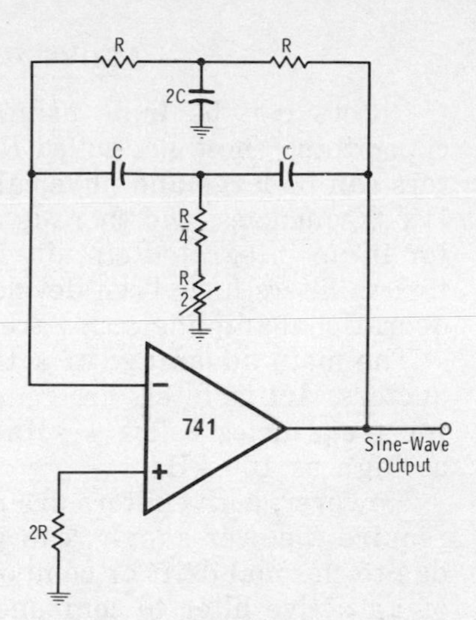

Fig. 5-21. Sine-wave oscillator.

A twin-T network is used in the feedback loop of this oscillator. When the twin-T network is slightly detuned, the phase of the feedback is shifted 180°, and the circuit will oscillate. The potentiometer in the circuit is used to detune the twin-T network by decreasing the resistance of the potentiometer until

Table 5-1. Some Standard Component Values for $f(\text{Hz}) = 1/(2\pi RC)$

$\frac{1}{2\pi RC}$ (Hz)	R (kΩ)	C (μF)
32	10.0	0.5
57	5.6	0.5
96	3.3	0.5
177	1.8	0.5
318	1.0	0.5
483	3.3	0.1
885	1.8	0.1
1590	1.0	0.1
2840	5.6	0.01
4830	3.3	0.01
8850	1.8	0.01
15,900	1.0	0.01

a sine wave appears at the output; if the resistance of the potentiometer is decreased too much, the sine wave will become distorted. The twin-T network is also very useful in certain active filter designs, which will be discussed in the next section.

ACTIVE FILTER DESIGN

Filters can be built using only resistors, inductors, and capacitors. These are called *RLC* or *passive* filters. But inductors can be large and physically awkward to use, especially at low frequencies, and there is no easy way to include an inductor in an integrated circuit. In recent years, however, inductorless filters have been developed. These new filters often use operational amplifiers and are called *active* filters.

The main advantage of active filters is that they use no inductors. Active filters are generally preferred to passive filters for frequencies below 1 kHz but can be used at frequencies as high as 100 kHz.

However, active filters are not without disadvantages: They require a power supply, can generate noise, and can oscillate due to thermal drift or component aging. Thus, the sensitivity of an active filter to component drift is an important design consideration. Highly selective filters are the most prone to oscillation, and so a very selective active filter must have a very low sensitivity to drift. In spite of these disadvantages, active filters are very popular for low frequencies. Several different ways an active filter can be designed will be discussed in this section.

Gyrator

Perhaps the most obvious way to design an active filter is to start with a classical RLC filter and use active circuits to simulate the inductors. A *gyrator* is a circuit that is commonly used in active filters to simulate an inductor. A gyrator can be built using two op amps, and active filters using gyrators can have a very low sensitivity to drift. The value of inductance simulated by the gyrator shown in Fig. 5-22 is given by the formula:

$$\text{SIMULATED INDUCTANCE} = \frac{R_1 \times R_2 \times R_4 \times C_1}{R_3}$$

where,

the inductance is in henrys,
the resistances are in ohms,
the capacitance is in farads.

Notice that the one end of the inductor simulated by the gyrator is grounded. It is difficult to use gyrators to simulate ungrounded, or "floating," inductors. Since most passive low-pass filters require floating inductors, gyrators are usually not used to build low-pass active filters. Instead, the *frequency-*

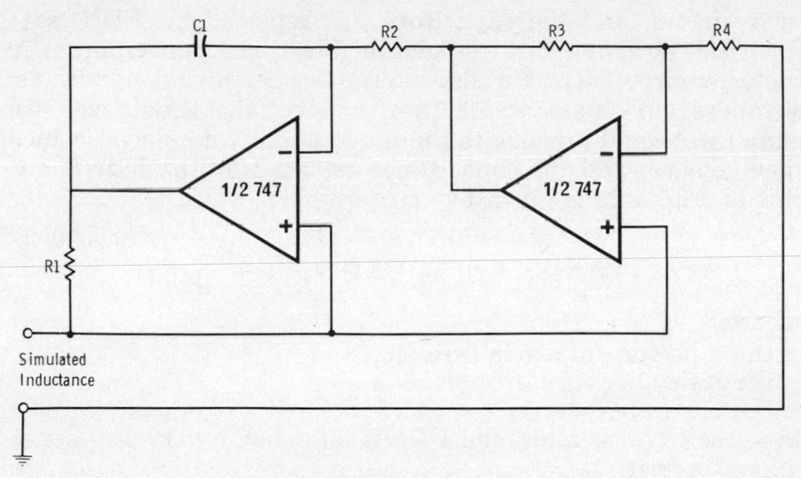

Fig. 5-22. Gyrator circuit.

dependent, negative-resistance circuit, or *FDNR,* can be used in low-pass active filters.

FDNR

Fig. 5-23 shows the circuit diagram of an FDNR filter. Active-filter designs using the FDNR circuit are based on a classical RLC low-pass filter. The resistors of the classical circuit are replaced by capacitors, the inductors are replaced

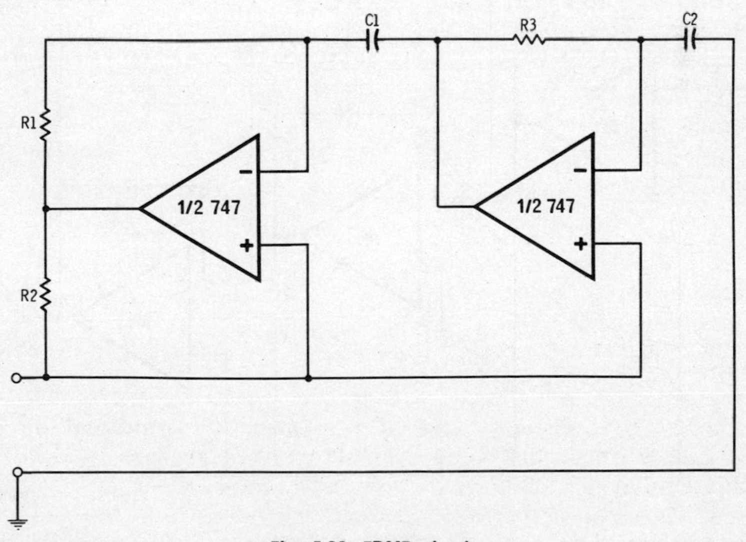

Fig. 5-23. FDNR circuit.

by resistors, and the capacitors are replaced by FDNRs to form the new, inductorless active filter. The capacitances in the new circuit (in farads) equal the reciprocal of the resistances (in ohms) which they replace; the resistances (in ohms) are equal in value to the inductances (in henrys) which they replace; and the capacitance replaced by the FDNR circuit of Fig. 5-23 is given by the formula:

$$\text{CAPACITANCE REPLACED} = \frac{R_1 \times R_3 \times C_1 \times C_2}{R_2}$$

where,

the capacitances are in farads,
the resistances are in ohms.

Like the gyrator filter, the FDNR filter has a very low sensitivity to drift.

Multiloop Feedback

Another approach to the design of active filters is to simulate the entire filter, using analog computer techniques, rather than just simulating the components of a classical RLC filter. A

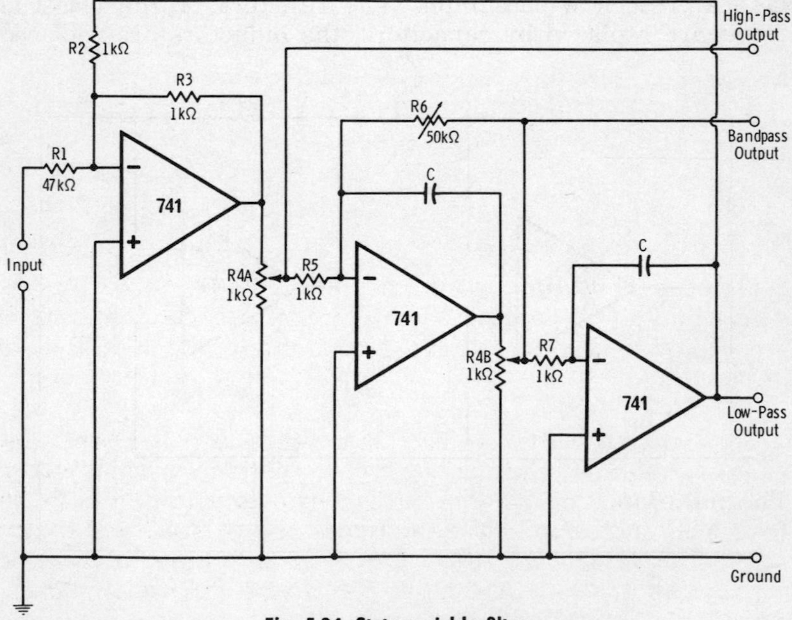

Fig. 5-24. State variable filter.

filter designed in this way is called a *multiloop feedback filter* or *state variable filter*.

Fig. 5-24 is the circuit diagram of a state variable filter. One advantage of this filter is that it can readily be used to perform different filtering functions by taking the output from different points; the output points for a high-pass, bandpass, and low-pass filter are shown in the diagram. Another advantage of the state variable filter is that the frequency and selectivity can easily be tuned independently; in Fig. 5-24, R_4 is used to adjust the frequency, and R_6 is used to adjust the selectivity. A third advantage is that this filter is relatively unaffected by the source and load impedances, so input and output buffer amplifiers usually are not needed. The disadvantages of the state variable filter are that it is more sensitive to component drift than the gyrator or FDNR filter, and more operational amplifiers are required. However, since buffer amplifiers are usually not needed, state variable filters may require fewer op amps for some applications.

Single-Loop Feedback

The simplest active filter is the *single-loop feedback filter*. This is a good general-purpose filter using just one op amp and one negative-feedback loop. The input component and/or feedback component have impedances that change with frequency. As the ratio of these impedances changes, the closed-loop gain of the circuit changes, and certain frequencies are either amplified or attenuated more than others.

The next several sections will give a variety of examples of active filters using IC op amps.

LOW-PASS FILTERS

A low-pass filter passes low-frequency signals but tends to reject signals above a certain cutoff frequency. Three low-pass active filters will be described: (1) a single-loop feedback filter, (2) a multiloop feedback filter, and (3) an FDNR filter.

Single-Loop Feedback

Fig. 5-25 shows the circuit of a single-loop feedback, low-pass active filter. This is a simple negative-feedback circuit. The impedance of the feedback component decreases as the frequency increases, while the impedance of the input component increases. The closed-loop gain of this circuit is therefore greatest at lowest frequencies. The cutoff frequency for this filter is given by the formula:

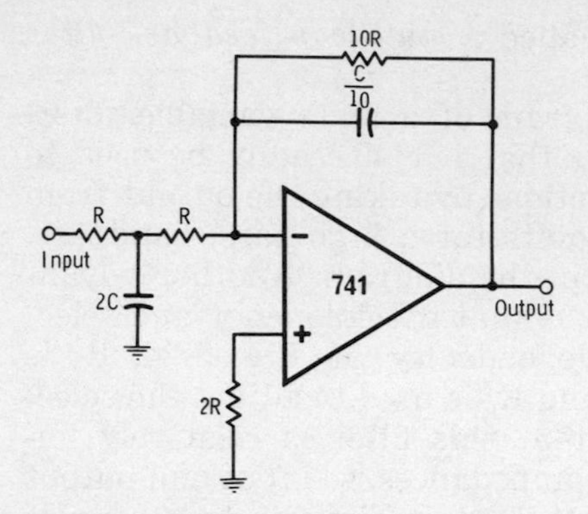

Fig. 5-25. Low-pass filter.

$$f(Hz) = \frac{1}{2\pi RC}$$

Some values for this formula are tabulated in Table 5-1.

If a very sharp frequency cutoff is required, however, a multiloop feedback filter or FDNR filter should be used.

Multiloop Feedback

A low-pass multiloop feedback filter can be built using the circuit of Fig. 5-24. The output is taken from the "low-pass output" terminal of the filter. The cutoff frequency of this filter is given by the formula:

$$f(Hz) = \frac{k}{2\pi \times R_4 \times C}$$

In this formula, C is in farads, R_4 is the total resistance of potentiometer R_4 in ohms, and k is a number that varies between 0 and 1 as the setting of potentiometer R_4 is changed (k is equal to 1 when there is maximum resistance between the potentiometer brush and ground). Table 5-1 lists some values for this formula with $k = 1$.

FDNR

The FDNR low-pass, active-filter design is always based on some classical RLC passive filter. For example, the classical "T-section" shown in Fig. 5-26A is a low-pass RLC filter with a cutoff frequency given by:

$$f(Hz) = \frac{1}{\pi\sqrt{2LC}}$$

Fig. 5-26 shows the equivalent FDNR active filter where resistors have replaced the inductors, a capacitor has replaced

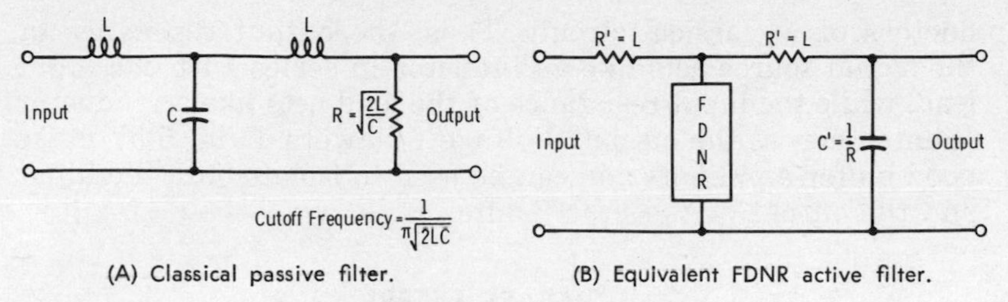

(A) Classical passive filter. (B) Equivalent FDNR active filter.

Fig. 5-26. FDNR filter design procedure.

the resistor, and an FDNR (see Fig. 5-23) has replaced the capacitor of the original circuit.

Fig. 5-27A is an example of a low-pass RLC filter circuit with a cutoff frequency of 320 Hz. The component values shown are not practical values for a real RLC filter, but the values do transform nicely to make the equivalent FDNR active filter shown in Fig. 5-27B. Fig. 5-27C shows the complete schematic diagram of the final FDNR active filter, using a 747 dual op amp.

When using the FDNR filter, it is important that the input signal source have a low output resistance, and that the output load have a high input resistance. The reason for this is that in the FDNR filter, the resistors behave like the in-

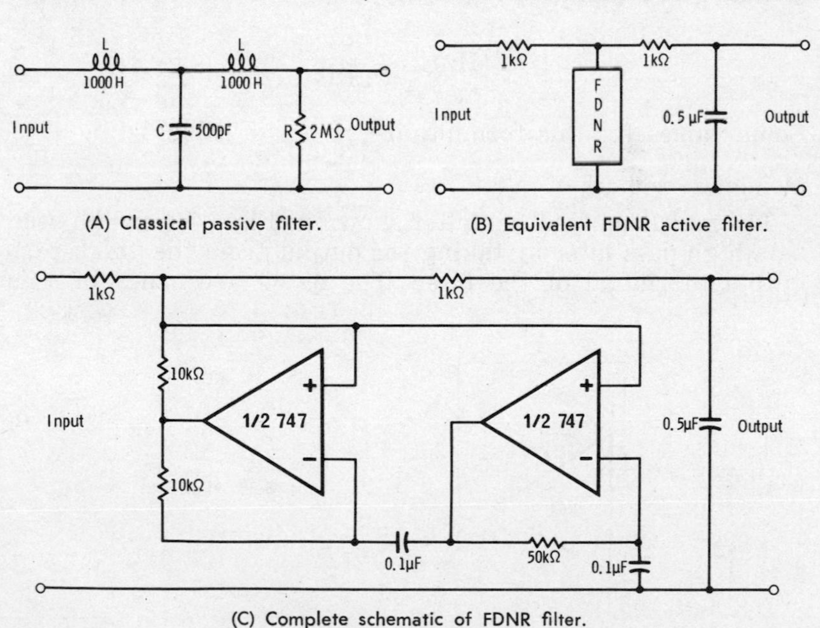

(A) Classical passive filter. (B) Equivalent FDNR active filter.

(C) Complete schematic of FDNR filter.

Fig. 5-27. FDNR low-pass filter with 320-Hz cutoff frequency.

ductors of a classical circuit. Thus, the output resistance of the signal source acts like an inductor in series with the input lead, while the input resistance of the load acts like an inductor shunted across the output. Voltage followers (Fig. 5-5) make good buffer amplifiers and can be used to isolate both the input and the output of the FDNR filter.

HIGH-PASS FILTERS

A high-pass filter tends to reject low-frequency signals while passing signals above a certain cutoff frequency. The three high-pass filters discussed in this section are the single-loop feedback filter, the multiloop feedback filter, and the gyrator filter.

Single-Loop Feedback

Fig. 5-28 is the circuit diagram of a single-loop feedback, high-pass active filter. The impedance of the input component decreases as the frequency increases, thus increasing the closed-loop gain of the circuit at higher frequencies. A resistor is included in series with the input lead to limit the gain at very high frequencies. This circuit tends to attenuate signals below the cutoff frequency given by:

$$f(\text{Hz}) = \frac{1}{2\pi RC}$$

Some values for this formula are given in Table 5-1.

Multiloop Feedback

The multiloop feedback filter shown in Fig. 5-24 can be used as a high-pass filter by taking the output from the "high-pass output" terminal of the filter. The cutoff frequency of this

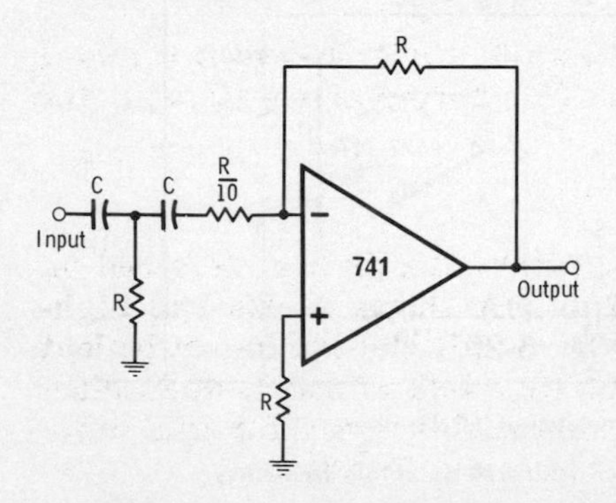

Fig. 5-28. High-pass filter.

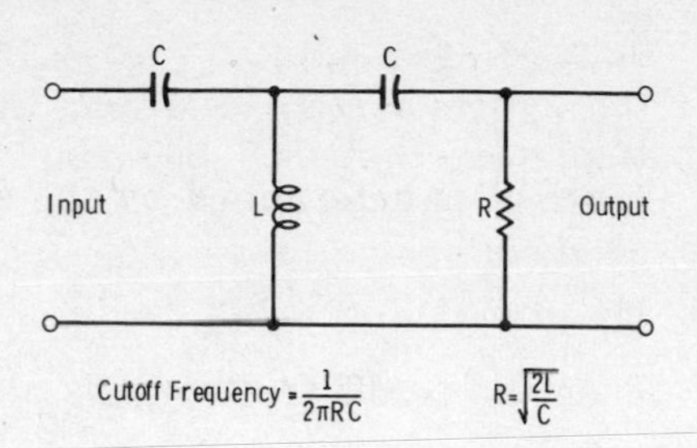

(A) Classical passive filter.

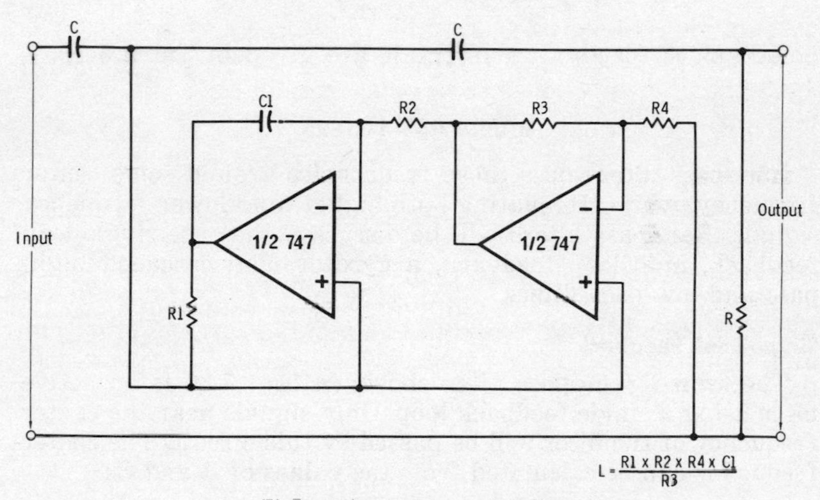

(B) Equivalent gyrator active filter.

Fig. 5-29. High-pass filter using gyrator.

filter is determined by the same formula as was used for the low-pass multiloop feedback filter.

$$f\,(\text{Hz}) = \frac{k}{2\pi \times R_4 \times C}$$

Again k, the potentiometer constant, varies between 0 and 1 as the setting of R_4 is varied. This formula is evaluated for $k = 1$ in Table 5-1.

Gyrator

Like the FDNR filter, the gyrator filter is always based on some classical RLC filter. Fig. 5-29A shows a classical high-pass, "T-section" filter, and Fig. 5-29B shows the equivalent active filter with the gyrator of Fig. 5-22 replacing the inductor of the original circuit. The cutoff frequency for this high-pass filter is given by:

$$f(\text{Hz}) = \frac{1}{2\pi\sqrt{2LC}}$$

When R is determined by the formula:

$$R = \sqrt{\frac{2L}{C}}$$

Then the cutoff frequency is also given by the formula:

$$f(\text{Hz}) = \frac{1}{2\pi RC}$$

Some values for this last formula are given in Table 5-1.

BANDPASS FILTERS

Bandpass filters pass those frequencies around some center frequency, while attenuating both higher- and lower-frequency signals. Bandpass filters will be discussed that use single-loop feedback, multiloop feedback, a gyrator, and cascaded high-pass and low-pass filters.

Single-Look Feedback

The twin-T bandpass filter shown in Fig. 5-30 is an active filter using a single feedback loop. Only signals near the center frequency of the filter will be passed by this circuit. The center frequency can be calculated from the values of R and C:

$$f(\text{Hz}) = \frac{1}{2\pi RC}$$

This formula is evaluated for several standard component values in Table 5-1.

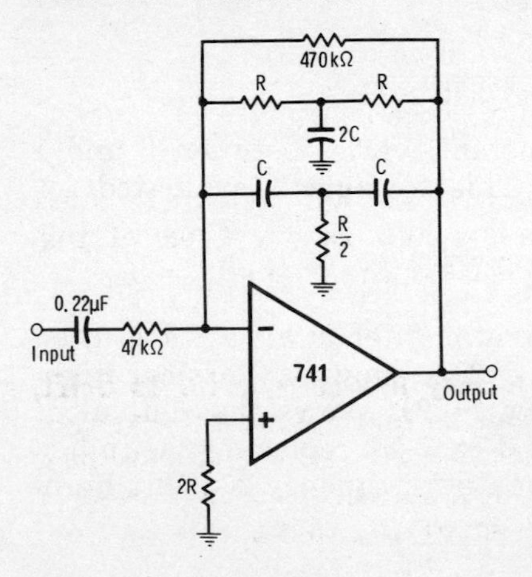

Fig. 5-30. Bandpass filter.

For this circuit to work best, the capacitor in the upper "T" should be exactly 2C, and the resistor in the lower "T" should be exactly ½R. If standard components with this relationship are not available, two resistors of value R can be put in parallel to yield a value of ½R. Similarly, two capacitors can be put in parallel to yield a capacitance of 2C.

The operation of the twin-T bandpass filter is straightforward. At the center frequency, the twin-T network in the feedback loop passes no current; this is because the current in the upper "T" is equal to and 180° out of phase with the current in the lower "T" at the −input of the op amp. For frequencies other than the center frequency, however, current does flow through the twin-T network. This means that the circuit will have maximum gain at the center frequency.

Multiloop Feedback

The multiloop feedback filter of Fig. 5-24 can be used as a bandpass filter by taking the output from the "bandpass output" terminal of the circuit. The selectivity, or "Q," of a bandpass filter is defined as:

$$Q = \frac{\text{BANDWIDTH}}{\text{CENTER FREQUENCY}}$$

The selectivity of this state variable filter is determined by the formula:

$$Q = \frac{R_6}{R_5}$$

when R_5 is equal to R_7. The circuit of Fig. 5-24 has a Q of 50 when R_6 is set at its maximum value. The center frequency of the filter is determined by the formula:

$$f(\text{Hz}) = \frac{k}{2\pi \times R_4 \times C}$$

Again k is the potentiometer constant of R_4, varying between 0 and 1 as the setting of R_4 is changed. Some values of this formula are tabulated in Table 5-1 for the case of $k = 1$.

Gyrator

Because gyrator filters have a very low sensitivity to drift, very selective bandpass filters can be built using the gyrator. Gyrator bandpass filters are based on classical RLC bandpass filters, and the design procedure is identical to that used to design the high-pass gyrator filter of Fig. 5-29.

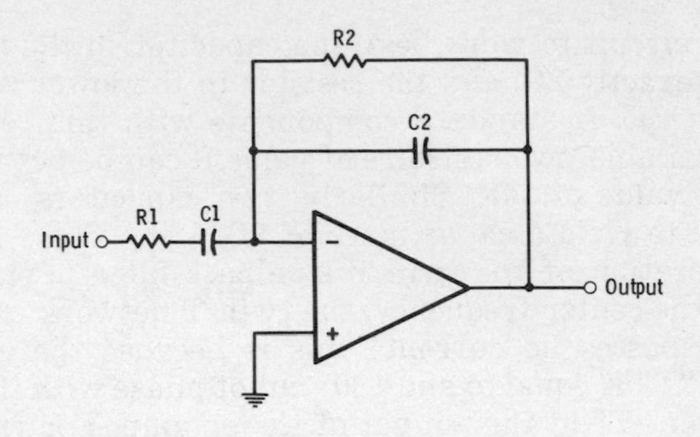

Fig. 5-31. Bandpass filter with independently adjustable high and low cutoff frequencies.

Cascaded Filters

Another way to build a bandpass filter is to cascade a low-pass and a high-pass filter. An example, using single-loop feedback, is shown in Fig. 5-31. The advantages of realizing a bandpass filter by the cascading method are that a very broad passband can be attained, and the high and low cutoff frequencies of the filter can be tuned separately. The cutoff frequency of the high-pass filter determines the low-frequency cutoff, and the cutoff frequency of the low-pass filter determines the high-frequency cutoff.

For the circuit of Fig. 5-31, the low-frequency cutoff is determined by R_1 and C_1 and can be found in Table 5-1. The high-frequency cutoff is determined by R_2 and C_2 and can also be found from Table 5-1. The narrowest passband is obtained when the product of R_1 and C_1 is equal to the product of R_2 and C_2.

NOTCH FILTERS

A notch filter passes all frequencies except those near its center frequency. The twin-T network used in the single-loop feedback bandpass filter can also be used to build a very effective notch filter. For a bandpass filter, the twin-T network is used in the feedback path, but for a notch filter, the twin-T is used in the input.

Fig. 5-32 is the circuit diagram of a twin-T notch filter. Since the same twin-T network is used as in the bandpass filter (Fig. 5-30), the same formula for center frequency obtains:

$$f\,(\mathrm{Hz}) = \frac{1}{2\pi RC}$$

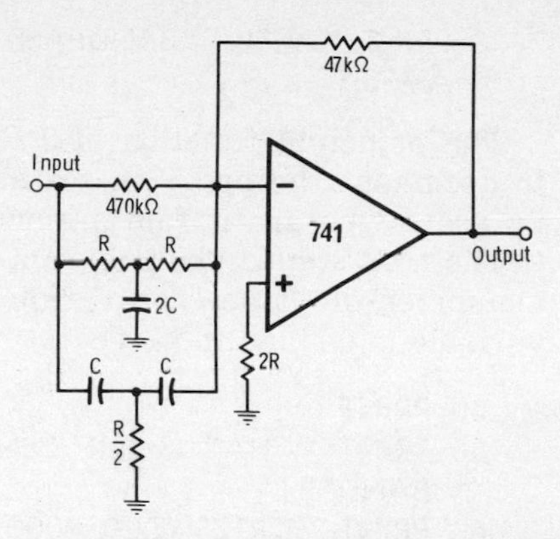

Fig. 5-32. Notch filter.

This formula is evaluated for several standard component values in Table 5-1. As for the twin-T bandpass filter, it is important that the components be well matched for best performance

COMPUTER-AIDED DESIGN

The design of more complicated filters using operational amplifiers can be greatly simplified by the use of a computer. Now that low-cost computers and programmable calculators are widely available, nearly any serious circuit designer is able to make use of computer-aided design techniques.

Take, for example, the bandpass filter circuit of Fig. 5-33. A computer program, written in BASIC, can be used to aid the designer in selecting the proper parts values for the filter.

The first three statement lines of the BASIC program are as follows:

```
10 PRINT
20 PRINT "THIS PROGRAM CALCULATES DATA NEEDED TO CON-
   STRUCT";
```

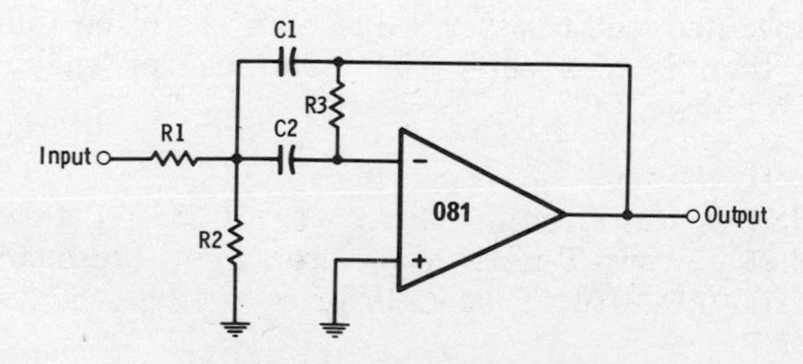

Fig. 5-33. Multiloop feedback bandpass filter.

```
30 PRINT "ACTIVE BANDPASS FILTERS USING 741-TYPE OP-
   AMPS."
```

The principal function of PRINT statements 20 and 30 is to document the program so that it can be run in the future by users who are unfamiliar with the program.

The next step in the program is to request the user to input a number of values for various parameters. Note that each variable is defined in preceding print statements.

```
40 PRINT
50 PRINT "WHAT IS THE CENTER FREQUENCY OF THE PASS
   BAND, IN";
60 PRINT "HERTZ? (E.G., 4000, 250, 60)"
70 PRINT
80 INPUT F
90 PRINT
100 PRINT "WHAT IS THE DESIRED GAIN, IN DECIBELS?"
110 PRINT "(E.G., 0, 5, 25)"
120 PRINT
130 INPUT H
140 PRINT
150 PRINT "WHAT IS THE DESIRED Q OF THE FILTER?"
160 PRINT
170 INPUT Q
180 PRINT
190 PRINT "SELECT A CONVENIENT STARTING VALUE C FOR
    CAPACITORS C1 AND";
200 PRINT "C2. IF THE VALUE IS IN PICOFARADS, ENTER THE
    DATA";
210 PRINT "IN THE FORMAT: X . . . E-12. IF THE VALUE IS IN
    MICROFARADS,";
220 PRINT "USE THE FORMAT: X . . . E-6."
230 PRINT
240 INPUT C
```

In this schematic formula, C_1 and C_2 are equal. Once a value for C is specified, the resistor values can easily be found. The next step then is to specify the formulas for each resistor value, as follows:

```
250 PRINT
260 PRINT
270 PRINT
280 PRINT "RESISTANCE IN OHMS"
290 PRINT
300 PRINT
```

```
310 LET W = F*2*3.14159
320 PRINT "R1=", Q/(H*W*C)
330 PRINT "R2=", Q/(((2*Q*Q) - H)*(W*C))
340 PRINT "R3=", (2*Q) / (W*C)
350 END
```

When the lines numbered 320, 330, and 340 are executed, the computer will print the proper resistance values for R_1, R_2, and R_3 according to the following formulas:

$$R_1 = \frac{Q}{2\pi fHC}$$

$$R_2 = \frac{Q}{(2Q^2 - H)\,2\pi fC}$$

$$R_3 = \frac{2Q}{2\pi fC}$$

where,

R is the resistance in ohms,
$\pi = 3.14159$,
f is the center frequency of the passband in hertz,
H is the gain of the passband center frequency in dB,
C is the capacitance in farads.

CHAPTER 6

Nonlinear Applications

An op amp is operating in its nonlinear range when the output of the op amp is not directly proportional to the input. Nonlinear operation occurs when the op-amp output reaches either its maximum possible excursion (positive saturation) or its minimum possible excursion (negative saturation).

IC op amps are used in their nonlinear range primarily for digital applications. The great advantage of using op amps rather than digital ICs is that op amps operating nonlinearly can be used in conjunction with op amps operating linearly; the input and output voltages are compatible, and the same power supply can be used for each circuit. For example, a triangle-wave generator can be made using two op amps, one operating nonlinearly, and one linearly. The first op amp is used as a comparator, and the second as an integrator. A triangle-wave generator of this type is called a *hysteresis oscillator,* as is discussed later in this chapter.

COMPARATOR

The comparator circuit of Fig. 6-1 is basic to all digital applications of operational amplifiers. A comparator simply compares an input voltage to a reference voltage. When the input signal is slightly greater than the reference voltage, the op amp swings into positive saturation. When the input is slightly less than the reference voltage, the op amp swings into

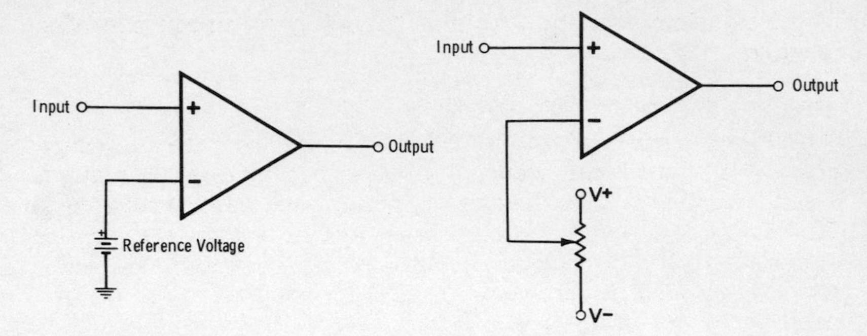

Fig. 6-1. Basic comparator circuit.

Fig. 6-2. Comparator with variable reference voltage.

negative saturation. The reference voltage can be made variable by using a potentiometer, as shown in Fig. 6-2.

If the output voltage of the comparator circuit is too large for a particular application, it can be limited by diodes. Fig. 6-3 shows how the output voltage of a comparator can be limited by two 1N914 diodes. When the comparator output exceeds 0.6 volt, one diode conducts and limits the output to that voltage. When the comparator output falls below −0.6 volt, the other diode conducts and similarly limits the output voltage.

When designing a comparator, two important parameters should be noted from the specification sheet (see Appendix). First, the op amp must have a sufficiently high maximum differential-input voltage so that the op amp will not be damaged by the maximal excursions of the input signal. Secondly, the op amp must have a fast slew rate if it is required to switch from saturation of one polarity to the opposite polarity very rapidly. Since compensation reduces the slew rate, fastest

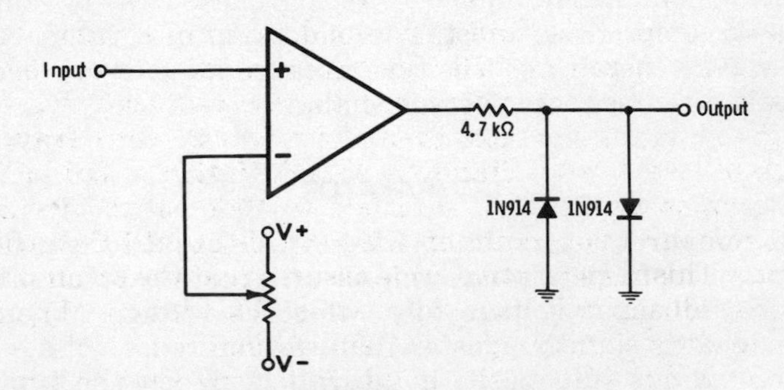

Fig. 6-3. Comparator with output limiter.

switching times can be attained by using an uncompensated op amp.

Sine-Wave to Square-Wave Converter

In Fig. 6-4, a comparator is shown being used to convert a sine wave to a square wave. The reference voltage is ground. When the sine wave goes positive, the op amp swings into positive saturation, and when the sine wave goes negative, the op amp swings into negative saturation. In this way, the comparator converts a sine wave to a square wave.

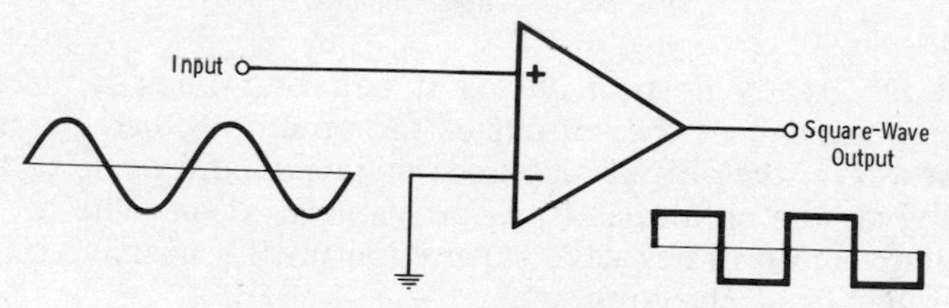

Fig. 6-4. Sine-wave to square-wave converter.

BISTABLE MULTIVIBRATOR

The bistable multivibrator, or "flip-flop," is a circuit with two possible stable output voltages, or "states." In the op-amp multivibrator, these two stable states are negative saturation and positive saturation. An input trigger signal is required to change the output from one state to the other.

Two different types of flip-flops can be built, depending on how the trigger signal is coupled to the bistable circuit. In the dc-coupled flip-flop the trigger signal is directly coupled, and in the ac-coupled flip-flop the trigger signal is capacitively coupled.

DC-Coupled Flip-Flop

Fig. 6-5 is the circuit diagram of a dc-coupled flip-flop. Since the feedback component (R_3) is between the output and the +input, this is a positive-feedback circuit. With positive feedback, if the output deviates even slightly from zero (say, due to noise inherent in all circuits), this deviation is fed back to the +input, is amplified, and causes further output deviation in the same direction as the initial deviation, until the op amp saturates. This regenerative cycle assures that the op amp with positive feedback will have only two stable states: (1) negative saturation and (2) positive saturation.

If the flip-flop of Fig. 6-5 is initially in positive saturation, a positive pulse applied to the input will have no effect—the

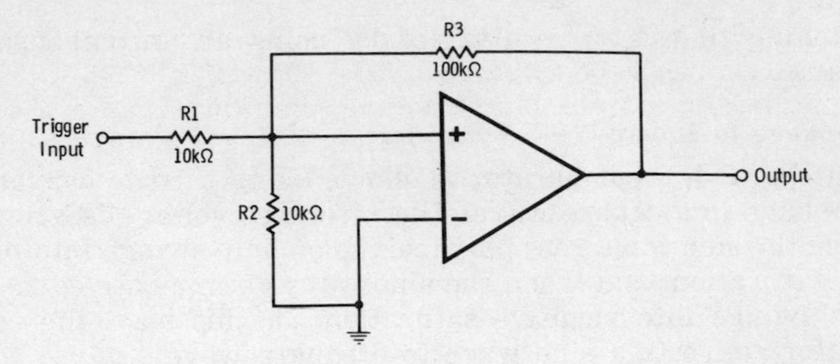

Fig. 6-5. Dc-coupled flip-flop.

output is already as positive as it can be. When a negative trigger pulse causes the +input of the op amp to become negative, however, the output also goes negative, and positive feedback drives the op amp into negative saturation. The op amp will then remain in negative saturation until a positive trigger pulse appears at the input.

In this dc-coupled flip-flop, two different signals are needed to switch the circuit back and forth between its two states. A positive pulse flips the circuit from negative to positive saturation, and a negative pulse flips it from positive to negative saturation. In the ac-coupled flip-flop, however, the same pulse can do both jobs.

AC-Coupled Flip-Flop

Fig. 6-6 is the circuit diagram of an ac-coupled flip-flop. The trigger signal is capacitively coupled to the −input of the op amp, rather than the +input, although either arrangement could be used.

Assume that the op amp is initially being held in positive saturation by positive feedback from the voltage divider of

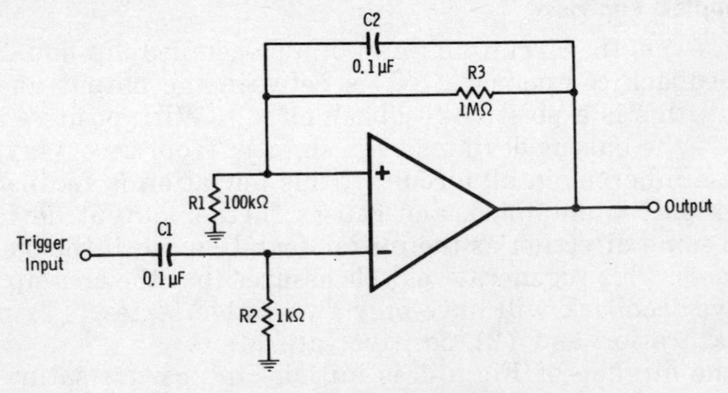

Fig. 6-6. Ac-coupled flip-flop.

R_3 and R_1. When the leading edge of a 2-volt positive trigger pulse is capacitively coupled to the −input through C_1, the −input becomes more positive than the +input, and the op amp swings into negative saturation. The op amp is held in negative saturation by positive feedback until the arrival of the next trigger pulse. When the trailing edge of the next trigger pulse is capacitively coupled to the −input, the −input becomes more negative than the +input, and the op amp swings back into positive saturation.

The purpose of C_2 is to assure that when the leading edge of a positive trigger pulse swings the op amp into negative saturation, the trailing edge of the same pulse will not immediately swing the op amp back into positive saturation. The value of C_2 shown is effective for pulses of less than 50-milliseconds duration, but longer trigger pulses require a larger capacitance.

ASTABLE MULTIVIBRATOR

Fig. 6-7 shows an op amp being used as an astable multivibrator, or square-wave generator. When the op amp is in positive saturation, the +input is held at one-half the positive saturation voltage by the voltage divider of R_1 and R_2. Capacitor C_1 is charging up through R_3 and R_4. But as soon as the capacitor voltage exceeds half the saturation voltage, the voltage at the −input becomes more positive than the voltage at the +input, and the op amp swings into negative saturation. The +input is now held at one-half the negative saturation voltage until C_1 charges to slightly less than half the negative saturation voltage; when this occurs, the op amp swings back into positive saturation, and the cycle repeats. A square wave is generated as the multivibrator circuit switches back and forth between its two astable states.

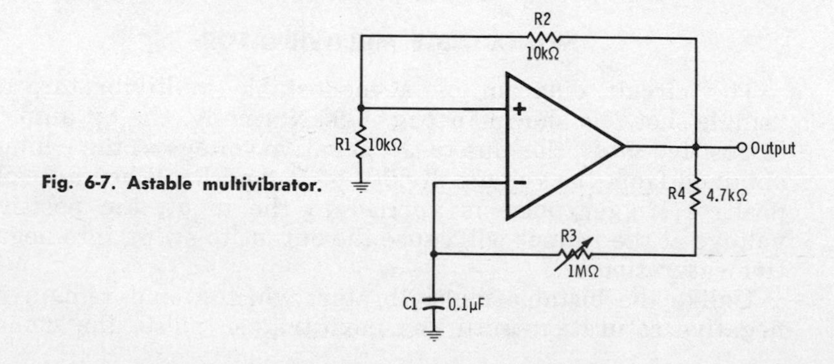

Fig. 6-7. Astable multivibrator.

The frequency of the square wave depends on the values of C_1, R_3, and R_4. Very low frequencies can be attained by using a large capacitance and large resistances. The high-frequency performance is limited by the slew rate of the particular op amp used. The exact frequency of oscillation is given by the formula:

$$f = \frac{1}{2.2C_1\ (R_3 + R_4)}$$

where,

f is the frequency in hertz,
the resistances are in ohms,
the capacitance is in farads.

A pictorial diagram of a 741 astable multivibrator designed to operate at 455 Hz is shown in Fig. 6-8.

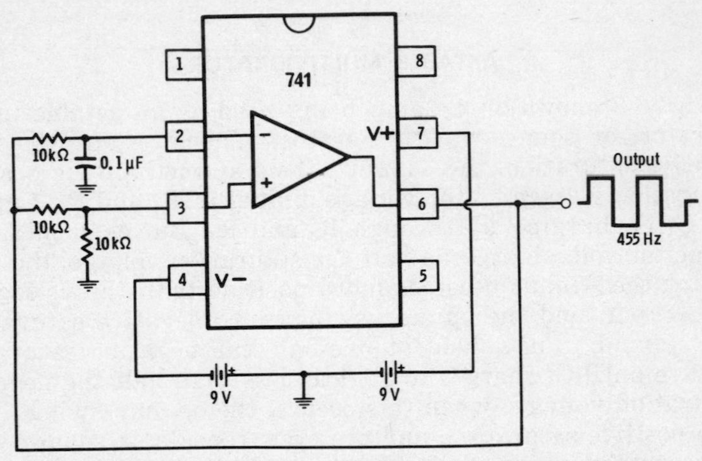

Fig. 6-8. Pictorial diagram of a 741 astable multivibrator designed to operate at 455 Hz.

MONOSTABLE MULTIVIBRATOR

The circuit diagram of a monostable multivibrator, or "single-shot," is shown in Fig. 6-9. Normally, the op amp is in positive saturation due to the negative voltage at the −input obtained from the voltage divider of R_2 and R_3. When a 2-volt positive trigger pulse is applied to the input, the positive voltage at the −input will cause the output to swing into negative saturation.

Unlike the bistable multivibrator, which would remain in negative saturation until the next trigger pulse, the mono-

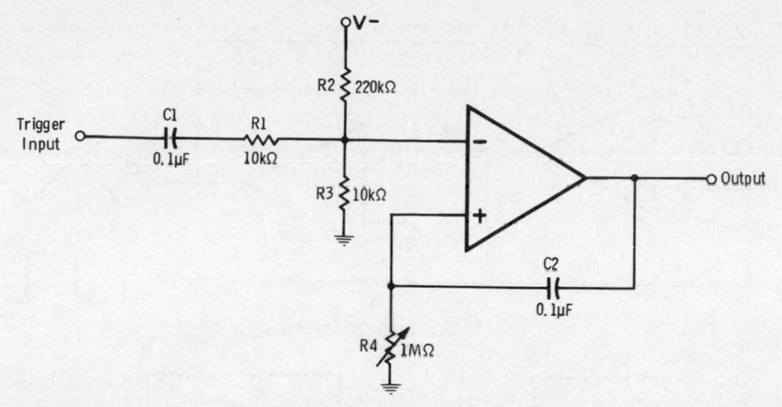

Fig. 6-9. Monostable multivibrator.

stable multivibrator automatically returns to positive saturation after a predetermined time interval. The length of this time interval is determined by C_2 and R_4. After the op amp swings into negative saturation, C_2 begins to charge up through R_4. As soon as the +input becomes more positive than the −input, the op amp returns to positive saturation.

The monostable multivibrator shown in Fig. 6-9 will remain in negative saturation for about 1 second with R_4 set at its maximum resistance, and for shorter durations when the setting of R_4 is reduced. Longer time periods can easily be attained by using a larger capacitance for C_2.

HYSTERESIS OSCILLATOR

An oscillator that produces both a triangle-wave and a square-wave output can be built using two operational amplifiers. One operational amplifier is used as a flip-flop, the other is used as an integrator. This circuit, shown in Fig. 6-10A, is called a *hysteresis oscillator*.

To understand how the hysteresis oscillator works, refer to the waveform diagram of Fig. 6-10B. Suppose that the output of the first operational amplifier (Output 1) is initially in positive saturation at +V volts. A current of V/R_2 will flow in resistor R_2 and this same current will charge the feedback capacitor, C_1. As the feedback capacitor charges, the output voltage of the second op amp (Output 2) will decrease linearly. As soon as Output 2 decreases to a voltage of $-V/2$, feedback through R_3 will cause Output 1 to swing to −V volts. Now Output 2 will increase linearly until a voltage of $+V/2$ is reached, at which point the cycle repeats. The exact fre-

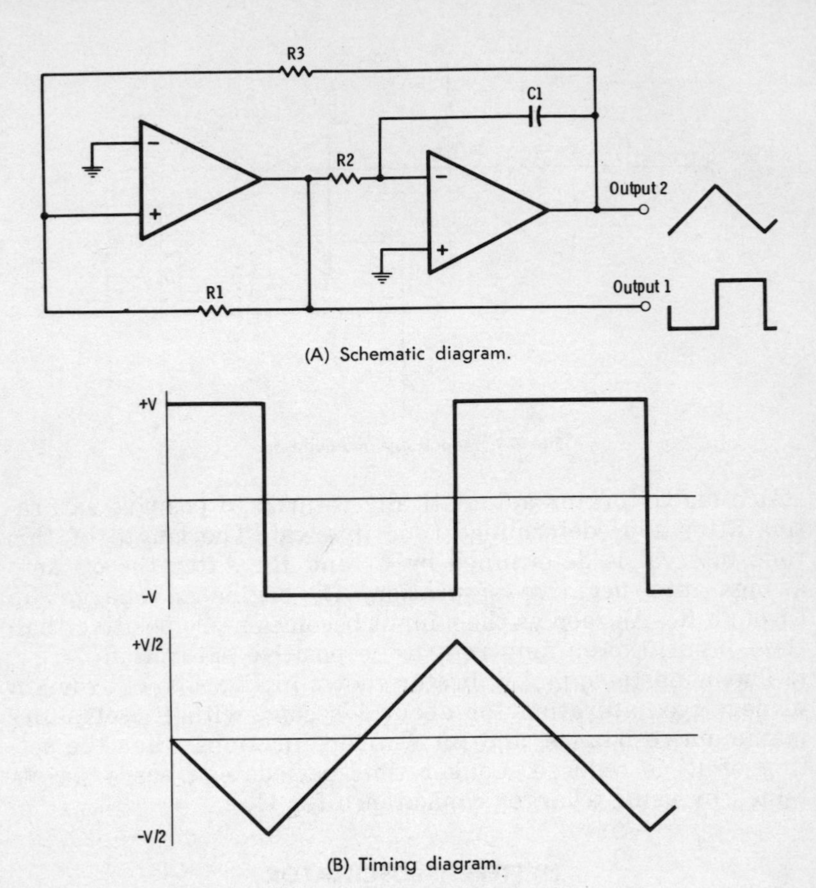

(A) Schematic diagram.

(B) Timing diagram.

Fig. 6-10. Hysteresis oscillator.

quency of oscillation for this circuit can be calculated from the following formula:

$$f = \frac{1}{2R_2C_1}$$

where,

f is the frequency in hertz,
the resistance is in ohms,
the capacitance is in farads.

SPECIAL-PURPOSE CIRCUITS

There are a great number of variations on the basic circuits presented in this chapter. With additional resistors, capacitors, and diodes, the basic comparator/multivibrator circuits

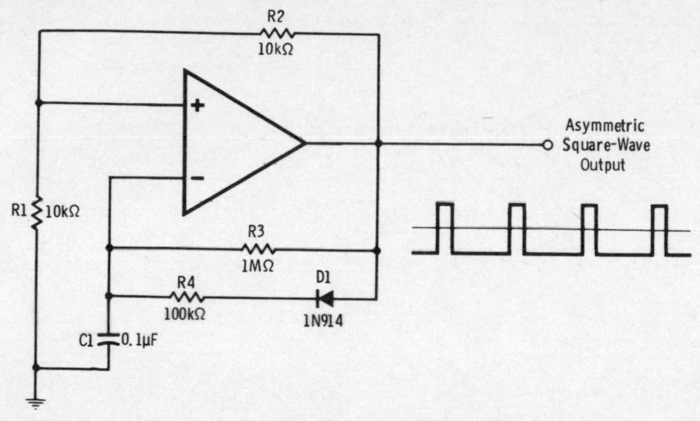

Fig. 6-11. Asymmetric square-wave generator.

can be transformed into a large number of special digital circuits.

To give an example of one such variation, consider the asymmetric square-wave generator circuit shown in Fig. 6-11. The circuit is basically that of an astable multivibrator (Fig. 6-7), but the addition of diode D_1 and resistor R_4 results in two different charging rates for capacitor C_1, depending on whether the op amp is in positive or negative saturation.

A great variety of op-amp circuits have been described in the last two chapters. Although these circuits are often useful by themselves, they are most useful when incorporated into larger electronic systems. The next, and final, chapter describes some electronic systems that have been developed using IC op-amp circuits.

CHAPTER 7

Op-Amp Systems

This chapter discusses the use of op-amp circuits in the design of electronic systems. Three circuits are used to illustrate the techniques involved. The first example, an audio EMG monitor, is an important application of an op-amp circuit in a medical-electronics monitoring system. The second example is an oscilloscope triggered-sweep generator, and the third example is a laboratory function generator.

AUDIO EMG MONITOR

When a muscle is contracted, a very small electrical voltage appears on the surface of the skin over the top of the muscle. This voltage is called the *electromyogram* or *EMG* of the muscle. As more force is developed by the muscle, the EMG voltage becomes larger, so that it is a good indicator of the force being generated by the muscle. The audio EMG monitor produces an audible note which varies in pitch as the EMG signal changes in amplitude.

Besides being an entertaining novelty, the audio EMG monitor has important medical applications. A stroke patient, for example, can use a portable audio EMG monitor to help him relearn to use muscle groups that were affected by the stroke. It is the IC op amp that has made it possible to build a small, lightweight, low-cost audio EMG monitor.

The block diagram of Fig. 7-1 shows the audio EMG monitor in three sections: (1) an IC op-amp differential amplifier, (2) a rectifier and averager, and (3) a current-controlled audio oscillator.

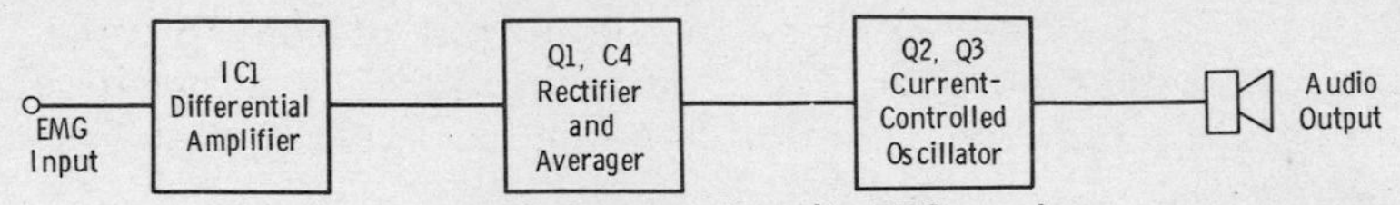

Fig. 7-1. Block diagram of audio EMG monitor.

Operation

The input signal to the EMG monitor comes from two electrodes which are taped to the skin over the muscle of interest. A small amount of electrode paste is used between the skin and the electrodes to assure a good electrical contact. The EMG signal is usually less than 1 millivolt in amplitude. A differential amplifier with good common-mode rejection must be used to amplify the EMG signal, in order to reject any 60-Hz noise picked up by the electrodes.

In the schematic diagram of the audio EMG monitor (Fig. 7-2), a type 741 op amp is shown used as a very high gain differential amplifier. This circuit is similar to that of Fig. 5-7, except that the feedback signal is taken from a voltage divider

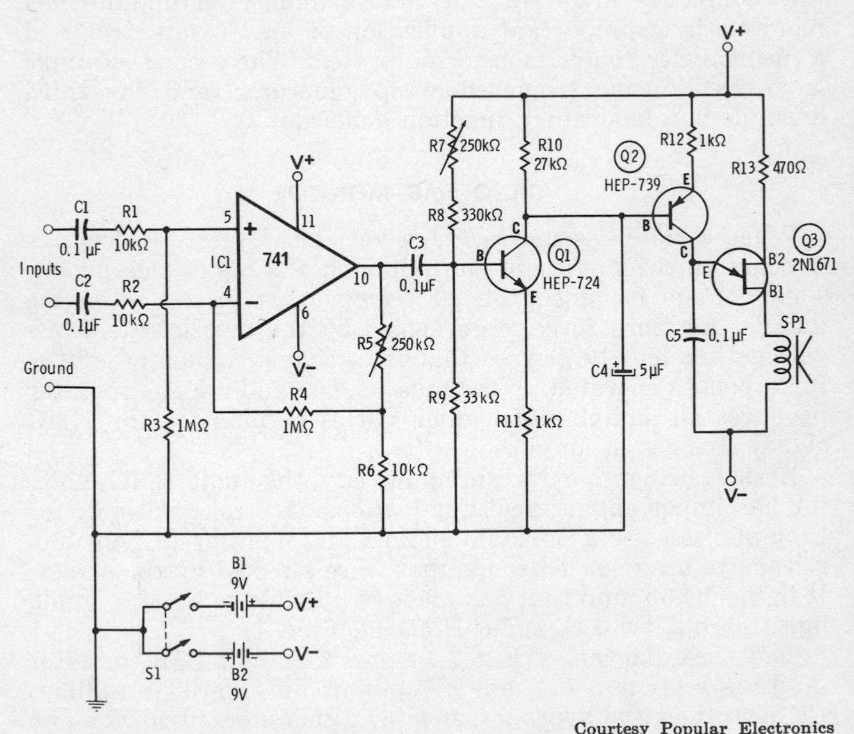

Courtesy Popular Electronics

Fig. 7-2. Schematic diagram of audio EMG monitor.

(R_5 and R_6), rather than from the output of the op amp, in order to reduce output offset due to input offset current. The gain of this differential amplifier can be varied from 100 to 2625 by the gain control, R_5.

In order to prevent changes in the output offset of the op amp from changing the biasing of the following stage, the output of the differential amplifier is capacitively coupled to the base of Q_1 through C_3. Transistor Q_1 acts as a half-wave rectifier by amplifying only the positive excursions of the op-amp output. Variable resistor R_7 is used to set the bias of Q_1 and serves as a threshold control. Capacitor C_4 is used to average the rectifier output of Q_1. The averaged, rectified, amplified EMG signal then controls the base current of Q_2.

Transistor Q_2, in turn, controls the charging current of C_5 and thus the oscillation frequency of the unijunction oscillator, Q_3. So the pitch of the tone produced by the speaker, which is in series with the base of the unijunction (Q_3), is controlled by the average magnitude of the original EMG signal.

It is important to note how the op-amp output and power supply are made compatible with the rest of this electronic EMG monitoring system. The output of the op amp is capacitively coupled to avoid problems with output offset. Transistor Q_1 is powered by only the V+ supply to effectively rectify the op-amp output. The unijunction oscillator, which requires a large supply voltage for best operation, is powered by both the V+ and V− supplies.

OSCILLOSCOPE TRIGGERED-SWEEP GENERATOR

Fig. 7-3 shows the block diagram of an oscilloscope triggered-sweep generator. The triggered-sweep generator uses three op amps, one linearly and two nonlinearly, and is a valuable addition to any oscilloscope that does not already have a triggered sweep. Usually, the signal applied to the vertical input of the oscilloscope is also connected to the input of the sweep generator, although an external triggering signal may also be used. The output of the triggered-sweep generator is connected to the external horizontal input of the oscilloscope.

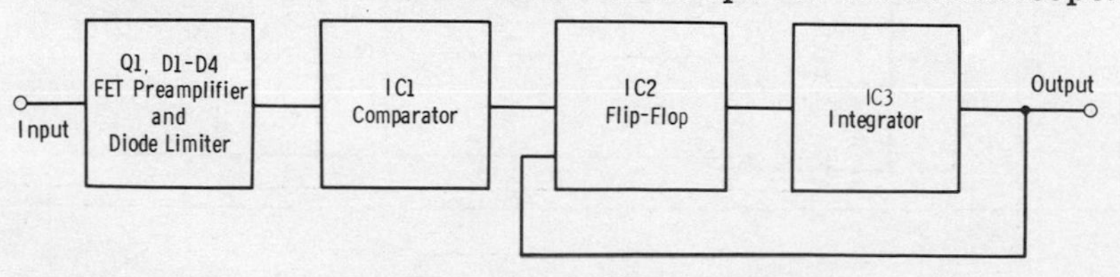

Fig. 7-3. Block diagram of triggered-sweep generator.

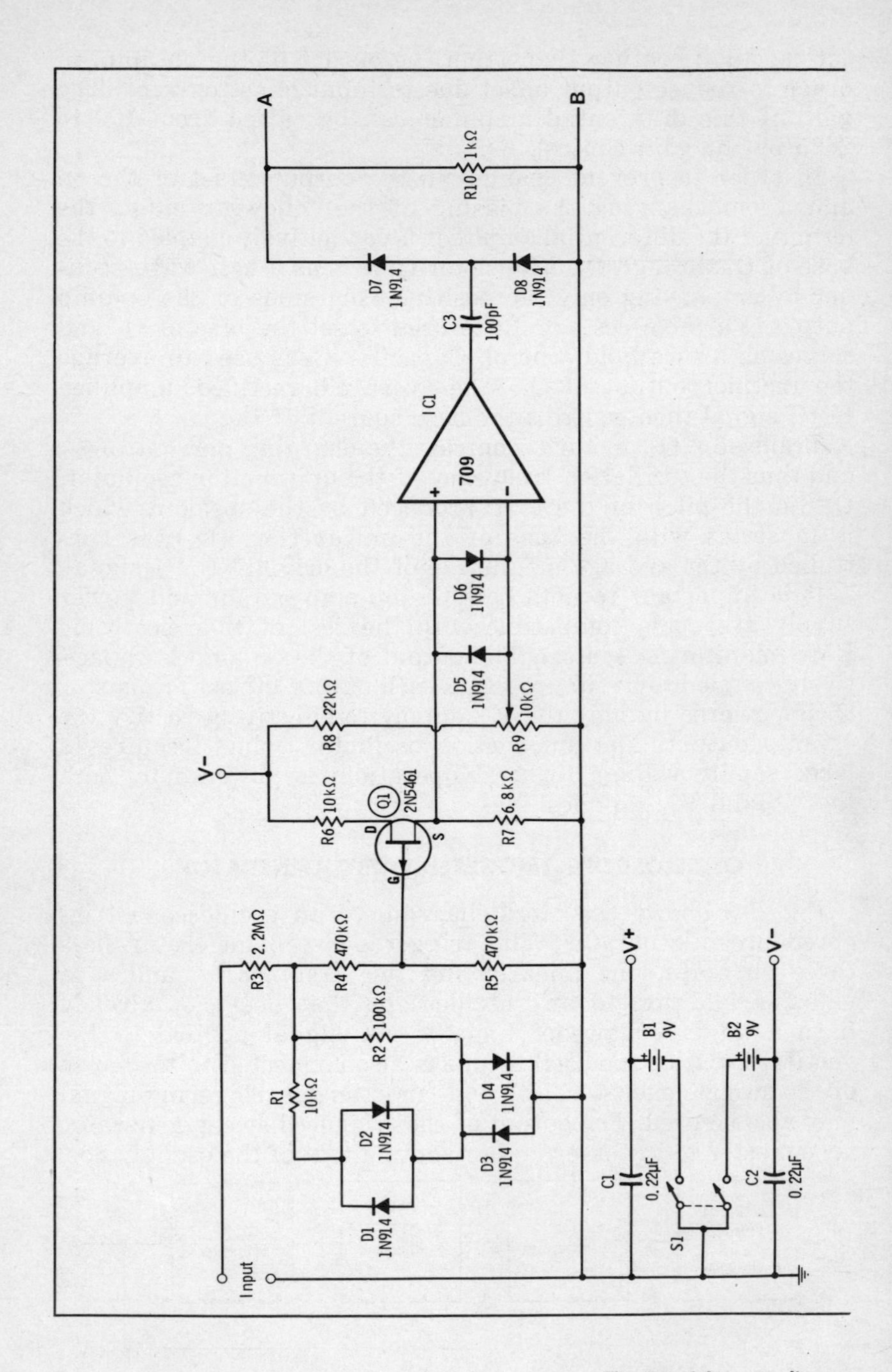

Fig. 7-4. Schematic diagram

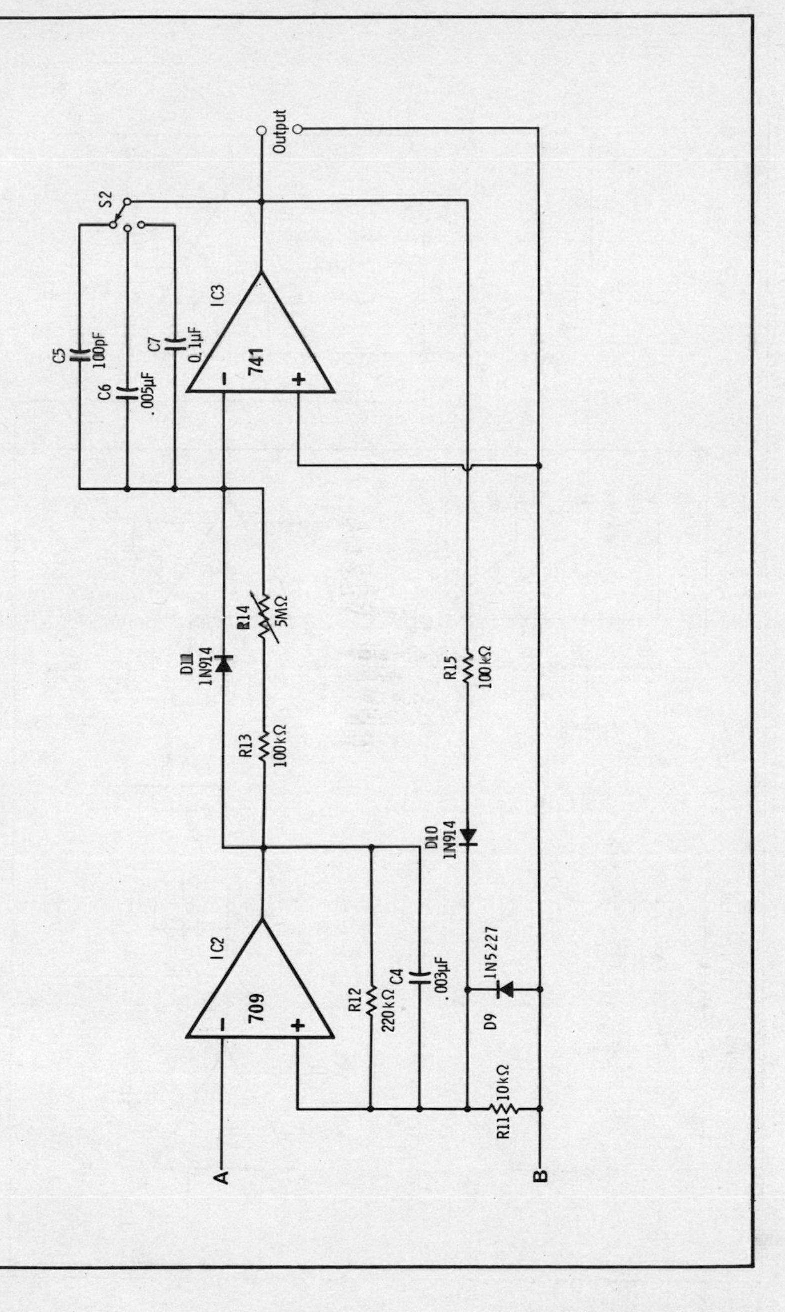

Courtesy Popular Electronics

of triggered-sweep generator.

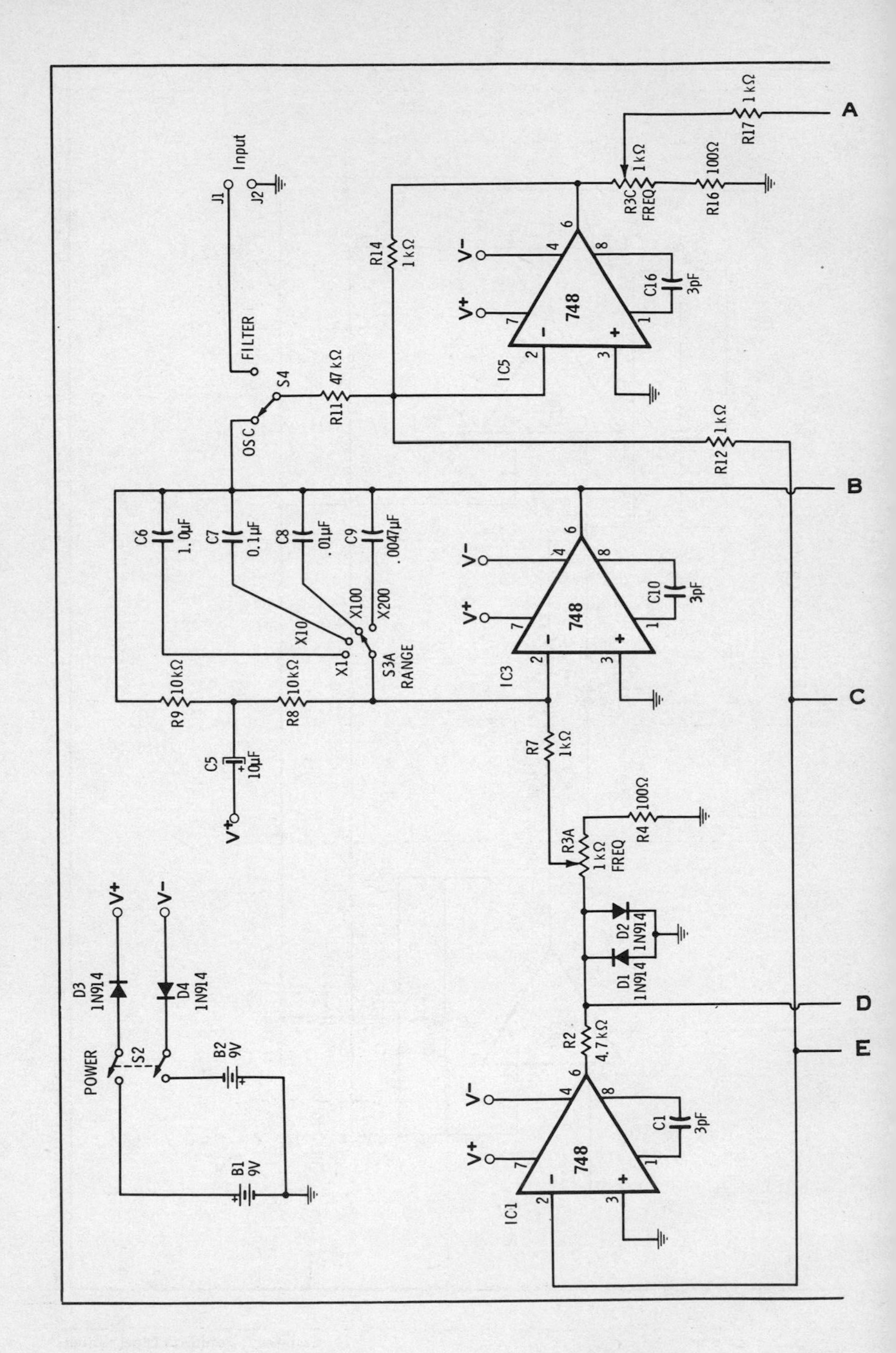

Fig. 7-5. Schematic diagram

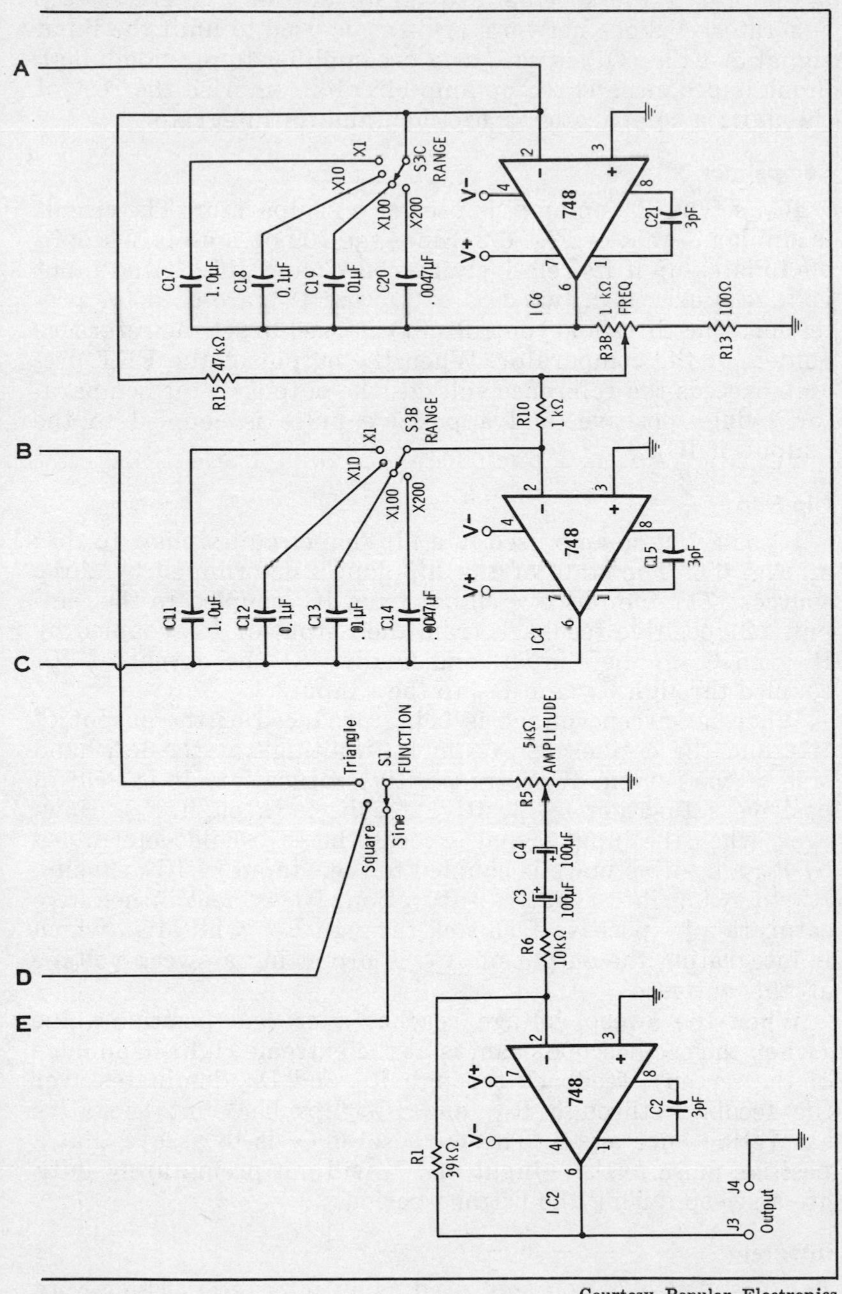

Courtesy Popular Electronics

of function generator.

Fig. 7-4 is the schematic diagram of the triggered-sweep generator. A diode network (D_1–D_4) is used to limit the input signal. A FET (Q_1) is used as a preamplifier to provide a high input impedance. Three op-amp circuits comprise the rest of the unit: a comparator, a flip-flop, and an integrator.

Comparator

IC_1, a type 709 op amp, is used as a comparator. The circuit is similar to that of Fig. 6-2. Since the 709 op amp is susceptible to latch-up if its relatively low maximum differential-input voltage is exceeded, two diodes (D_5 and D_6) are used for protection. The threshold control (R_9) is used to set the reference voltage of the comparator. When the output of the FET preamp exceeds the reference voltage, the output of the comparator swings positive, and a positive pulse is coupled to the −input of IC_2.

Flip-Flop

IC_2 is a 709 op amp used in a flip-flop circuit similar to that of Fig. 6-6. The state of the flip-flop is determined by three sources: (1) the positive signal from IC_1 coupled to the −input, (2) positive feedback from the output of IC_2 coupled by R_{12} and C_4 to the +input, and finally (3) the output of IC_3 coupled through R_{15} and D_{10} to the +input.

When no sweep voltage is being produced at the output of IC_3, and the oscilloscope beam is "waiting" at the left-hand side of the screen, IC_2 is in positive saturation. It is held in positive saturation by positive feedback through R_{12}. However, when the input signal exceeds the threshold determined by R_9, a positive pulse is coupled to the −input of IC_2, causing IC_2 to swing into negative saturation. IC_2 is held in negative saturation by positive feedback through R_{12}, while IC_3, which is integrating the output of IC_2, is producing a sweep voltage at the output.

When the sweep voltage reaches a critical positive value (when the oscilloscope beam is at the extreme right-hand edge of the screen), feedback through R_{15} and D_{10} dominates over the feedback through R_{12}, and IC_2 flips back into negative saturation once again. The purpose of C_4 is to assure that a positive pulse to the −input of IC_2 will not prematurely initiate a sweep during the retrace period.

Integrator

IC_3 is a type 741 op amp used as an integrator. The circuit is similar to that of Fig. 1-7A, except that a diode (D_{11}) is connected across the input resistance of R_{13} and R_{14}. The charging current of the feedback capacitor (C_5, C_6, or C_7) is dependent on whether D_{11} is forward or reverse biased. When IC_2 is in negative saturation, D_{11} is reverse biased, and the feedback capacitor charging current is determined by the setting of R_{14}. IC_3 integrates the negative output of IC_2 to produce a very linear sweep voltage with a sweep rate determined by the setting of R_{14} and S_2.

When IC_2 swings into positive saturation, D_{11} is forward biased, and the feedback capacitor rapidly discharges. As the capacitor discharges, the oscilloscope beam rapidly retraces from the right- to the left-hand side of the screen, where it remains until IC_2 again swings into negative saturation to initiate another sweep cycle. If desired, the output of IC_2 may be used to provide Z-axis blanking during retrace.

FUNCTION GENERATOR

Fig. 7-5 is the schematic diagram of a laboratory function generator which generates a low-distortion sine wave, square wave, and triangle wave. This function generator can be tuned over the entire audio spectrum in four ranges. Six type 748 IC op amps are the only active components used.

Although the schematic diagram may at first appear formidable, the block diagram of Fig. 7-6 reveals that the operation of the function generator can be understood in terms of circuits that have already been discussed in detail. IC_1, for example, is used as a comparator to convert a sine wave to a

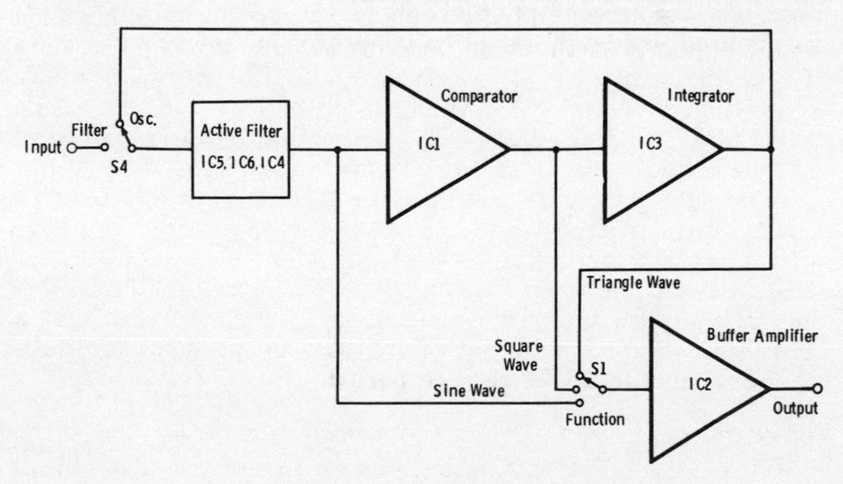

Fig. 7-6. Block diagram of function generator.

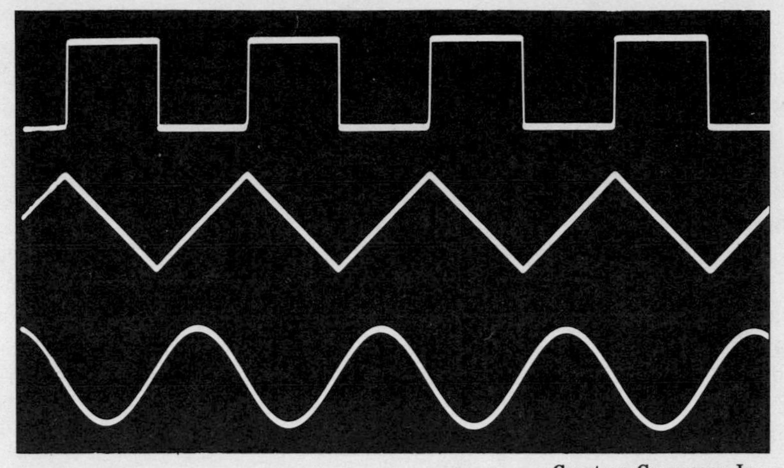

Courtesy Cromemco Inc.

Fig. 7-7. Output waveforms of function generator.

square wave, just as in the circuit of Fig. 6-4. IC_3 is used as an integrator to convert the square-wave output of IC_1 to a triangle wave, as discussed in Chapter 1. IC_4, IC_5, and IC_6 are used in a state variable filter, similar to that of Fig. 5-24, to filter out the fundamental sine-wave component of the triangle wave. IC_2 is used as a simple inverting amplifier.

Switch S_1 is used to select either the sine wave, square wave, or triangle wave as the output waveform, and S_4 is provided to select an optional input signal if the active filter section of the function generator is desired to be used for some special purpose. An oscilloscope photograph of the actual output waveforms produced by the function generator is shown in Fig. 7-7.

APPENDIX

Typical Op-Amp Specification Sheets

In this appendix, the various op amps are listed by their generic numbers. The manufacturers' numbers for these op amps are usually prefixed to identify the manufacturer. For example, the μA741, LM741, MC1741, and SN72741 are all type 741 op amps. In addition, these numbers are usually suffixed to specify the temperature range and package style.

In these specification sheets, base diagrams of the ICs are shown from a top view. Slew-rate specifications are given assuming unity-gain compensation.

Type: 080/081

Typical specifications for ±15-volt supply:

Gain	200,000 (50,000 minimum)
Input Offset Voltage	15 mV (maximum)
Input Offset Current	0.2 nA (maximum)
Input Bias Current	0.4 nA (maximum)
Common-Mode Rejection	70 dB (minimum)
Slew Rate	12 V/μs

Remarks:

1. The 081 is internally compensated. The 080 is compensated by a single external capacitor.
2. BIFET inputs are used for very low input bias current.
3. The 080 has same pin configuration as other uncompensated op amps such as the 101 and 748.
4. The 081 has same pin configuration as other compensated op amps such as the 741 and 3140.
5. Output short-circuit protected.
6. The 071 is a low-noise version of the 081 designed for audio applications.
7. The 061 is a low-power version of the 081 designed for battery-powered applications.

CONNECTION DIAGRAMS

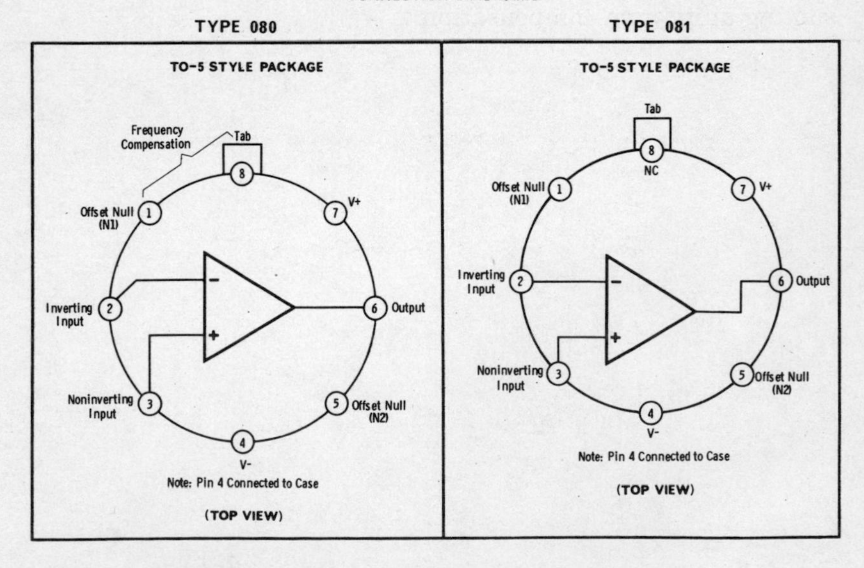

CONNECTION DIAGRAMS

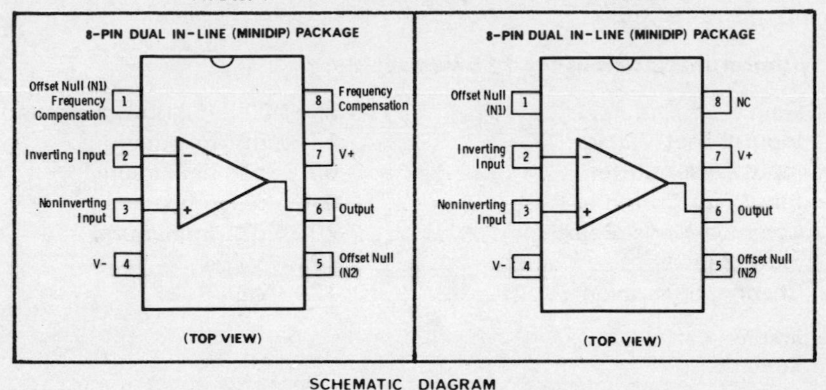

SCHEMATIC DIAGRAM

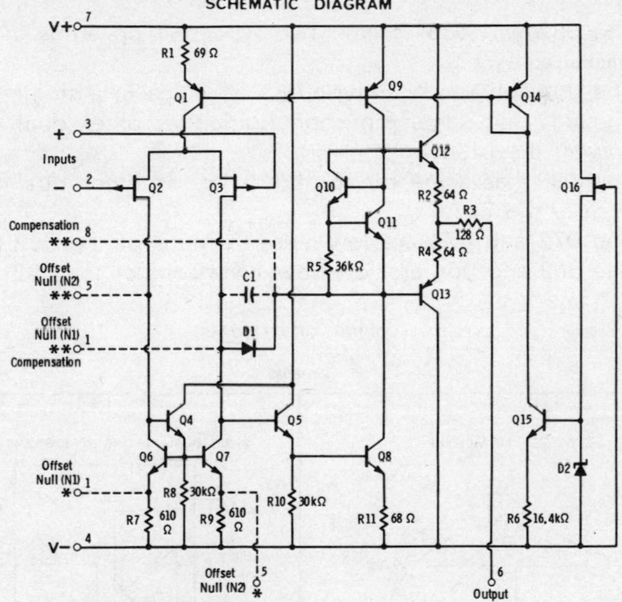

* Offset null control connections for type 081.
** Offset null control and frequency compensation connections for type 080.

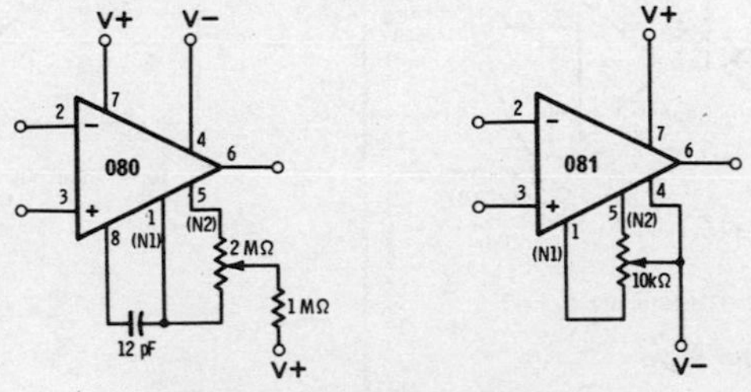

FREQUENCY COMPENSATION AND OPTIONAL OFFSET NULL CIRCUITS

Type: 082/083/084

Typical specifications for ±15-volt supply:

Gain	200,000	(50,000 minimum)
Input Offset Voltage	15	mV (maximum)
Input Offset Current	0.2	nA (maximum)
Input Bias Current	0.4	nA (maximum)
Common-Mode Rejection	70	dB (minimum)
Slew Rate	12	V/μs
Channel Separation	120	dB

Remarks:

1. The 082 and 083 contain two type 081 op amps in a single package.
2. The 084 contains four type 081 op amps in a single package.
3. The 082 has same pin configuration as other dual op amps such as the 1458.
4. The 083 has same pin configuration as other dual op amps such as the 747.
5. The 072 and 074 are low-noise versions of the 082 and 084.
6. The 062 and 064 are low-power versions of the 082 and 084.

CONNECTION DIAGRAMS

TYPE 082

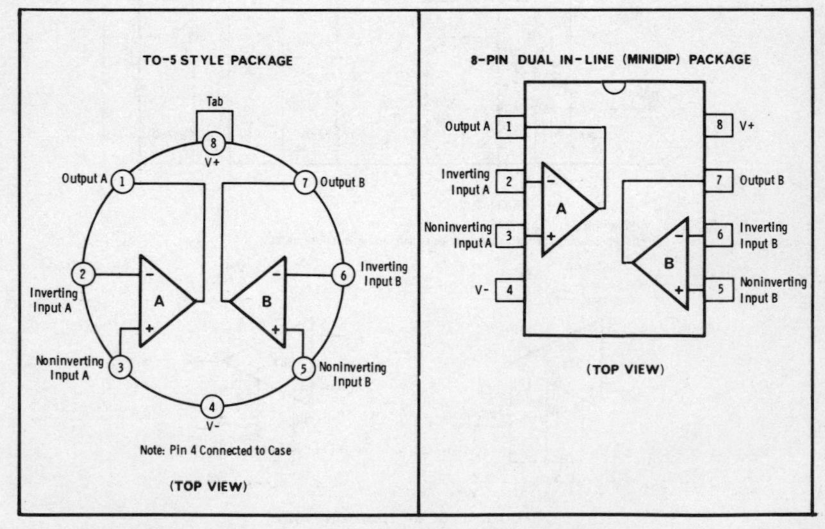

CONNECTION DIAGRAMS

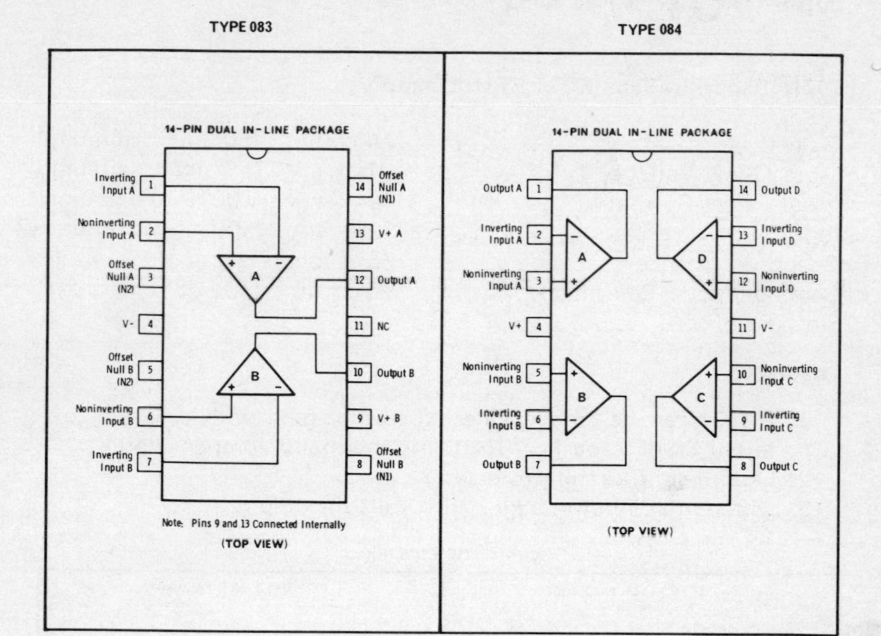

Type: 101

Typical specifications for ±15-volt supply:

Gain	160,000		(50,000 minimum)
Input Offset Voltage	1.0	mV	(5.0 mV maximum)
Input Offset Current	40	nA	(200 nA maximum)
Input Bias Current	120	nA	(500 nA maximum)
Output Resistance	75	ohms	
Common-Mode Rejection	90	dB	(70 dB minimum)
Slew Rate	0.5	V/μs	

Remarks:

1. LM101 can be compensated for unity gain with a single, external 30-pF capacitor. LH101 is internally compensated.
2. Output short-circuit protected.
3. Differential input voltage: ±30 volts maximum.

CONNECTION DIAGRAMS

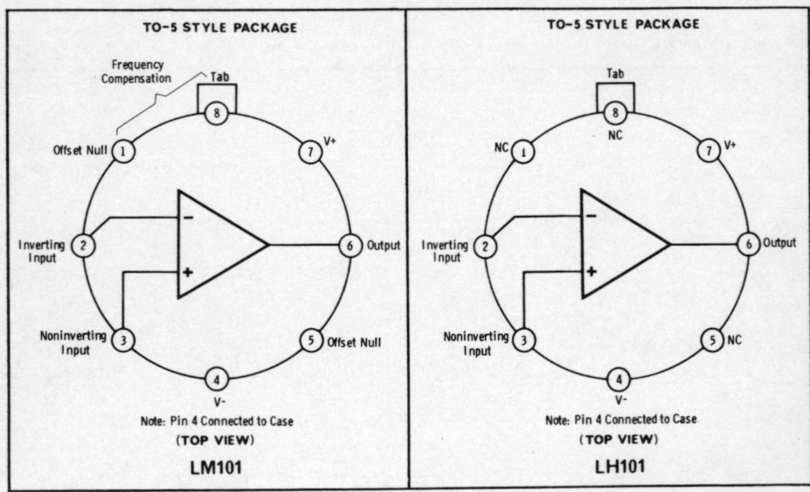

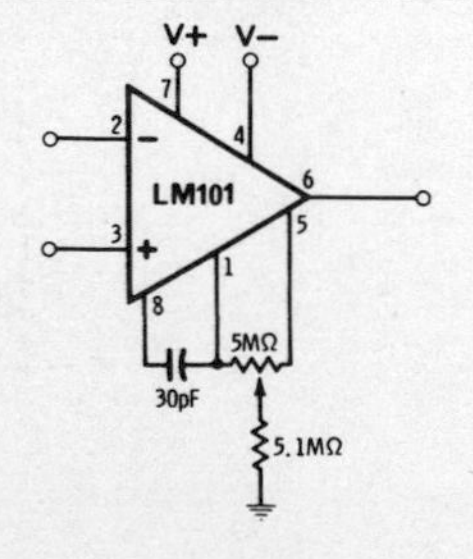

FREQUENCY COMPENSATION AND OPTIONAL OFFSET NULL CIRCUIT FOR LM101

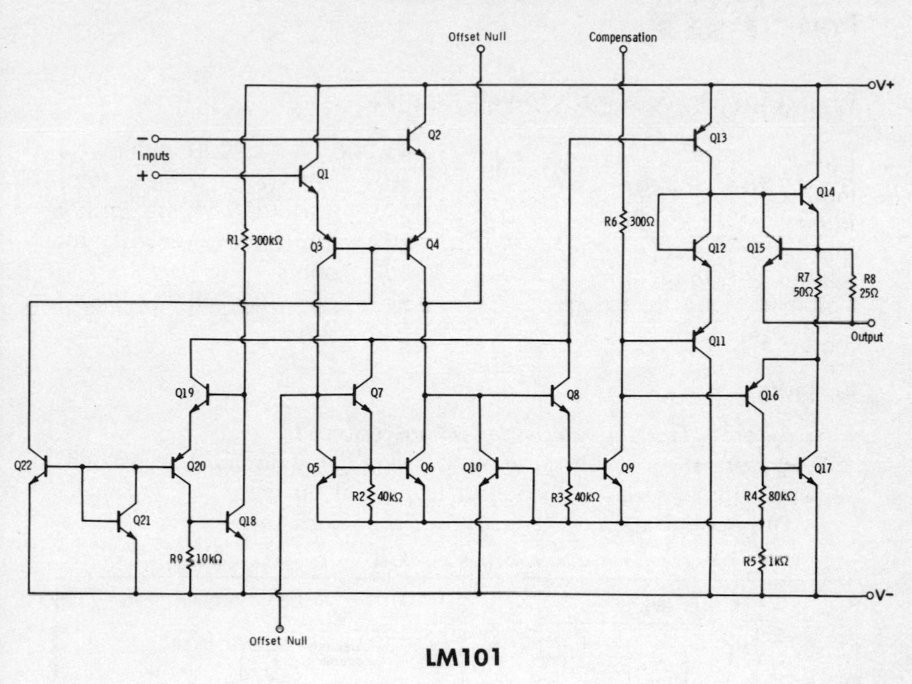
Offset Null
Compensation
V+
Inputs
Q1
Q2
Q3
Q4
Q5
Q6
Q7
Q8
Q9
Q10
Q11
Q12
Q13
Q14
Q15
Q16
Q17
Q18
Q19
Q20
Q21
Q22
R1 300kΩ
R2 40kΩ
R3 40kΩ
R4 80kΩ
R5 1kΩ
R6 300Ω
R7 50Ω
R8 25Ω
R9 10kΩ
Output
V−
Offset Null

LM101

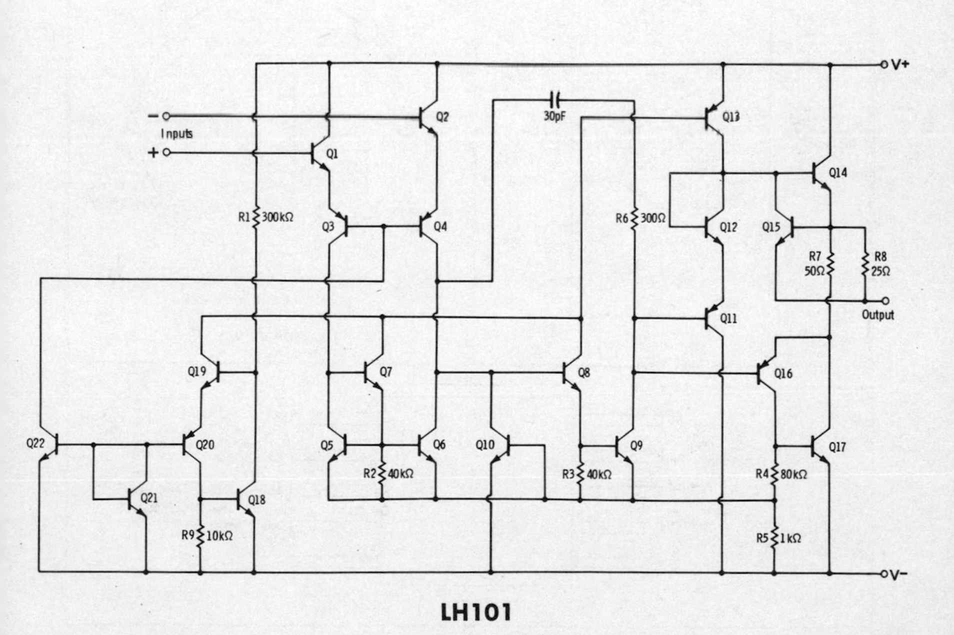
V+
Inputs
30pF
Q1
Q2
Q3
Q4
Q5
Q6
Q7
Q8
Q9
Q10
Q11
Q12
Q13
Q14
Q15
Q16
Q17
Q18
Q19
Q20
Q21
Q22
R1 300kΩ
R2 40kΩ
R3 40kΩ
R4 80kΩ
R5 1kΩ
R6 300Ω
R7 50Ω
R8 25Ω
R9 10kΩ
Output
V−

LH101

Type: 709

Typical specifications for ±15-volt supply:

Gain	45,000		(15,000 minimum)
Input Offset Voltage	1.0	mV	(7.5 mV maximum)
Input Offset Current	50	nA	(500 nA maximum)
Input Bias Current	200	nA	(1500 nA maximum)
Output Resistance	150	ohms	
Common-Mode Rejection	90	dB	(65 dB minimum)
Slew Rate	0.25	V/μs	

Remarks:

1. External frequency compensation required.
2. Output short-circuit duration: 5 seconds maximum. Use 50-ohm resistor in series with output to protect op amp.
3. Differential input voltage: ±5 volts maximum.

CONNECTION DIAGRAMS

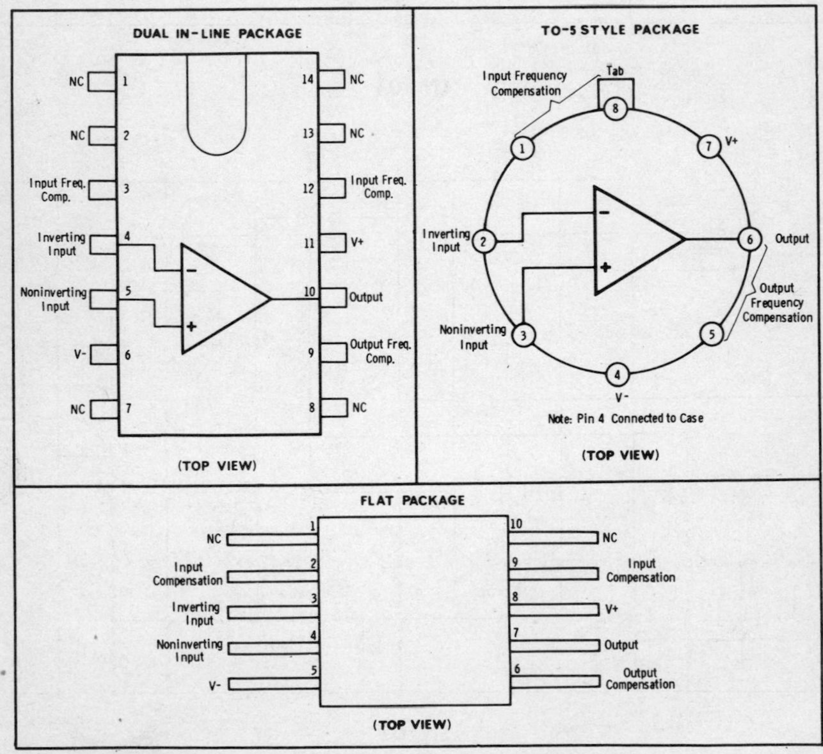

FREQUENCY COMPENSATION

Pin numbers shown
for TO-5 style package.

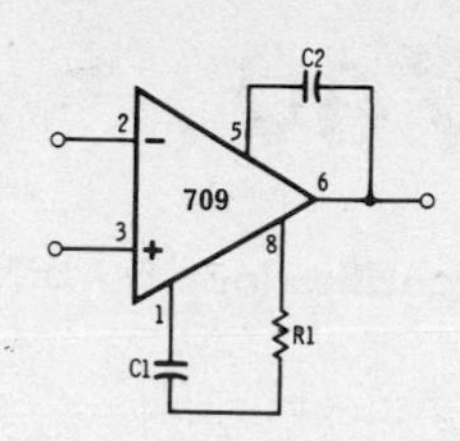

Closed-loop gain	Compensation components required: C1	R1	C2
1	5000 pF	1.5 kΩ	200 pF
10	500 pF	1.5 kΩ	20 pF
100	100 pF	1.5 kΩ	3 pF
1000	10 pF	0	3 pF

SCHEMATIC DIAGRAM

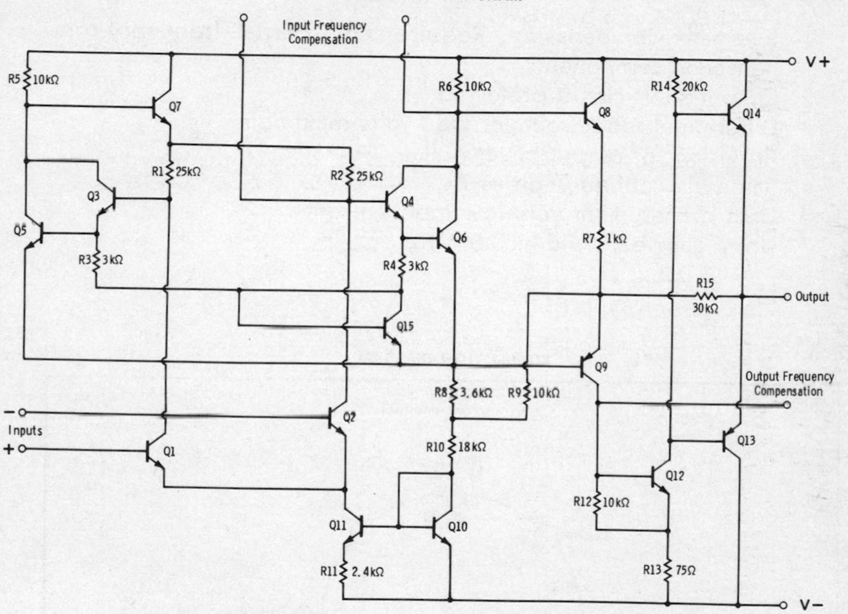

FREQUENCY RESPONSE FOR VARIOUS CLOSED-LOOP GAINS

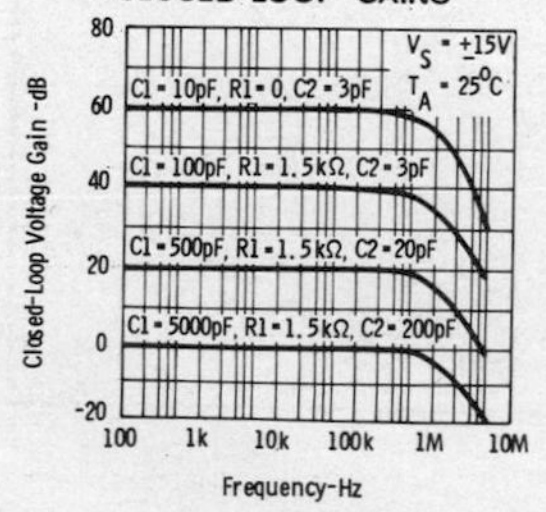

OUTPUT VOLTAGE SWING AS A FUNCTION OF FREQUENCY FOR VARIOUS COMPENSATION NETWORKS

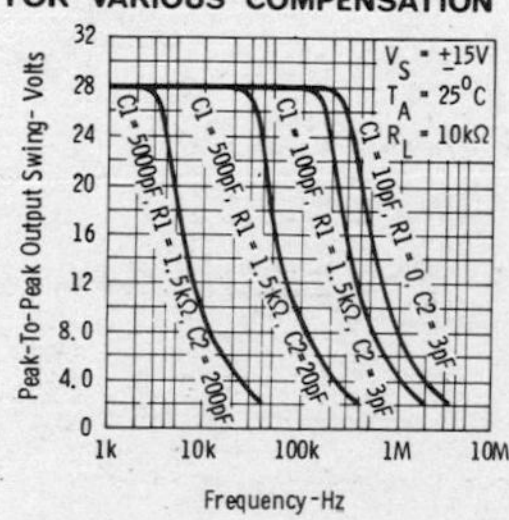

Type: 740

Typical specifications for ±15-volt supply:

Gain	1,000,000	(50,000 minimum)
Input Offset Voltage	10 mV	(20 mV maximum)
Input Offset Current	0.04 nA	(0.15 nA maximum)
Input Bias Current	0.1 nA	(0.2 nA maximum)
Output Resistance	75 ohms	
Common-Mode Rejection	80 dB	(64 dB minimum)
Slew Rate	6.0 V/μs	

Remarks:

1. Internally compensated. Requires no external frequency-compensation components.
2. Output short-circuit protected.
3. Differential input voltage: ±30 volts maximum.
4. Provision for output offset null.
5. Same pin configuration as 741.
6. Uses FET input for very low input currents.
7. Unity gain bandwidth: 3.0 Mhz.

CONNECTION DIAGRAM

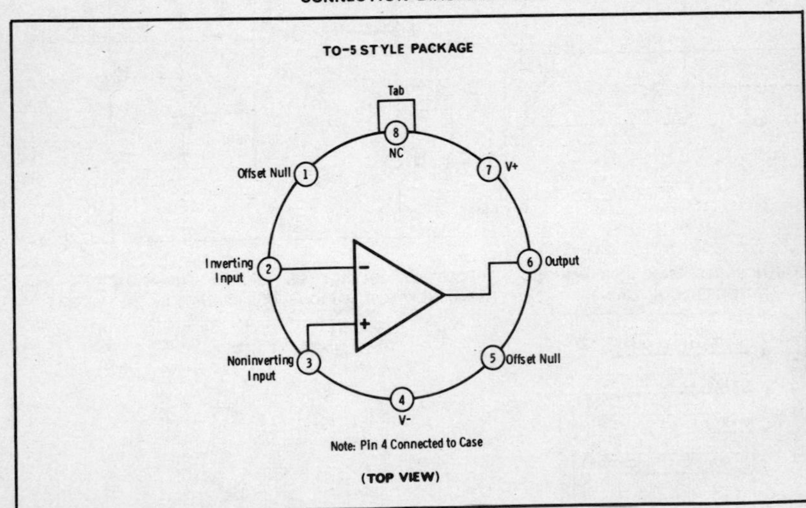

OPTIONAL OFFSET NULL CIRCUIT

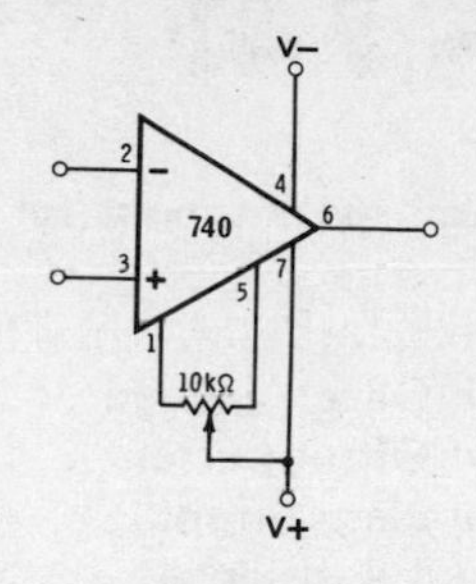

SCHEMATIC DIAGRAM

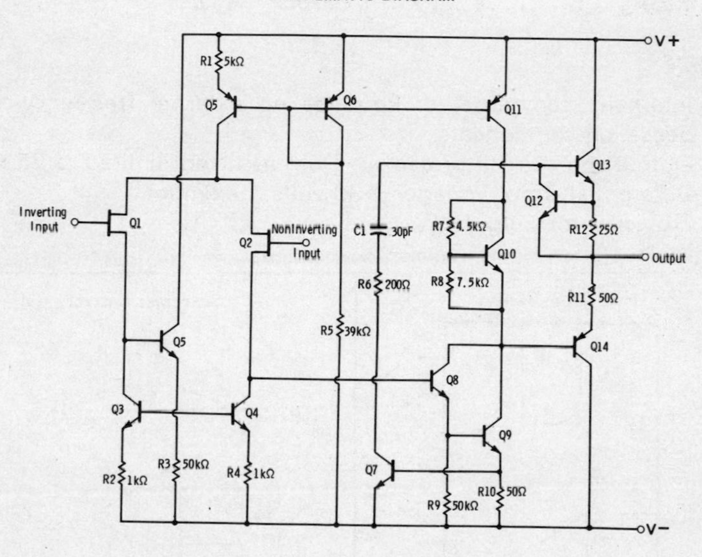

Type: 741

Typical specifications for ±15-volt supply:

Gain	200,000		(50,000 minimum)
Input Offset Voltage	1.0	mV	(5.0 mV maximum)
Input Offset Current	20	nA	(200 nA maximum)
Input Bias Current	80	nA	(500 nA maximum)
Output Resistance	75	ohms	
Common-Mode Rejection	90	dB	(70 dB minimum)
Slew Rate	0.5	V/μs	

Remarks:

1. Internally compensated. Requires no external frequency-compensation components.
2. Output short-circuit protected. Output current limited to 25 mA.
3. Differential input voltage: ±30 volts maximum.
4. Provision for output offset null.

CONNECTION DIAGRAMS

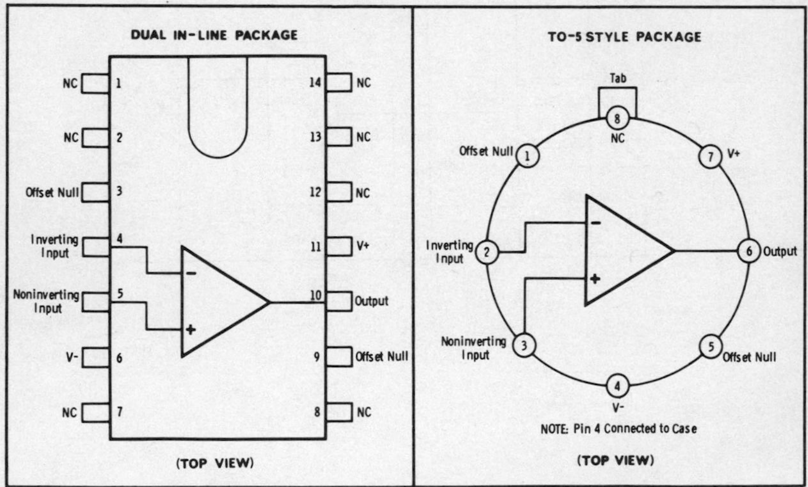

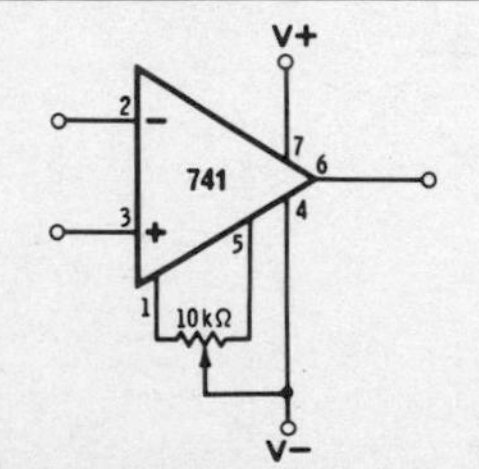

OPTIONAL OFFSET NULL CIRCUIT

Pin numbers shown for TO-5 style package.

SCHEMATIC DIAGRAM

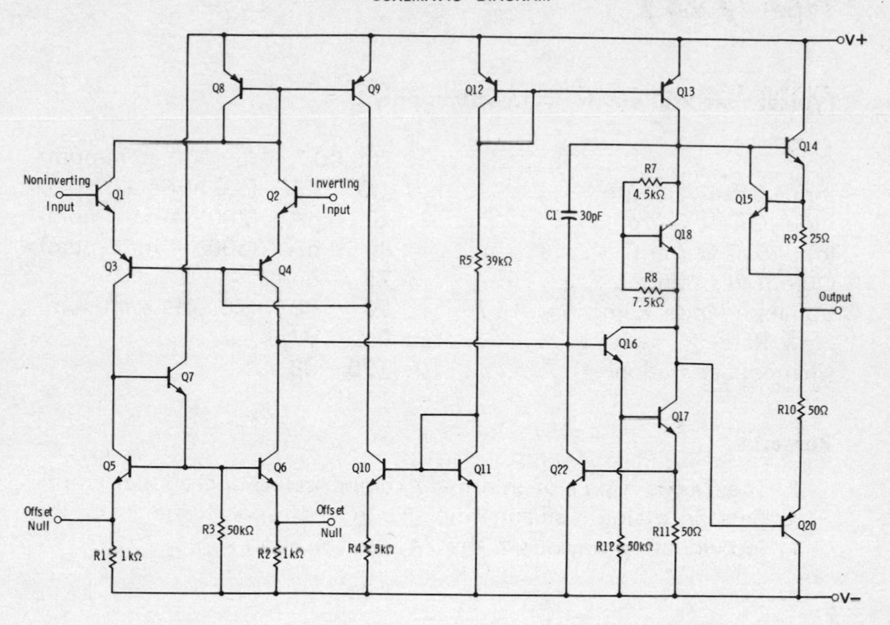

OPEN-LOOP VOLTAGE GAIN AS A FUNCTION OF FREQUENCY

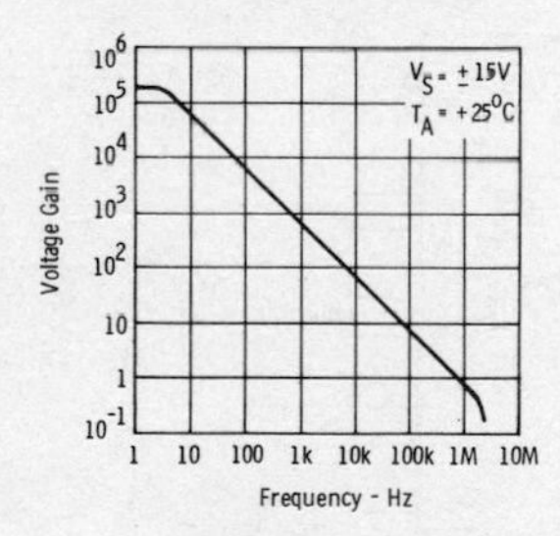

OUTPUT VOLTAGE SWING AS A FUNCTION OF FREQUENCY

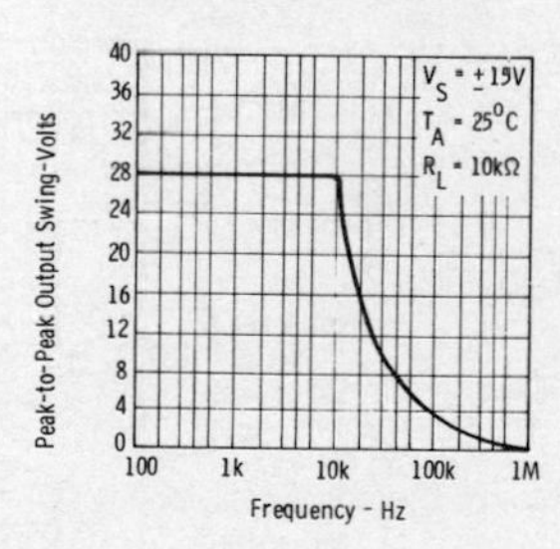

Type: 747

Typical specifications for ±15-volt supply:

Gain	200,000		(50,000 minimum)
Input Offset Voltage	1.0	mV	(5.0 mV maximum)
Input Offset Current	20	nA	(200 nA maximum)
Input Bias Current	80	nA	(500 nA maximum)
Output Resistance	75	ohms	
Common-Mode Rejection	90	dB	(70 dB minimum)
Slew Rate	0.5	V/μs	
Channel Separation	120	dB	

Remarks:

1. The 747 is a pair of internally compensated operational amplifiers on a single silicon chip. For further specifications for the individual op amps, see the 741 specification sheet.

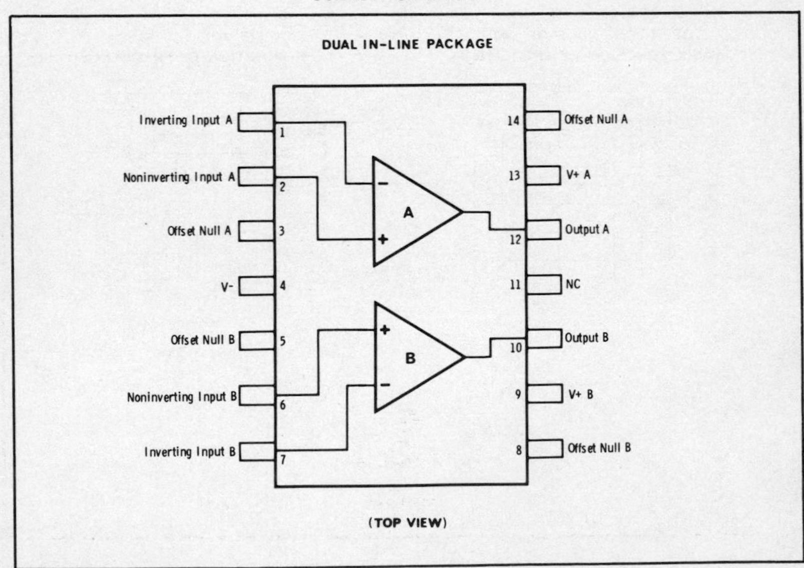

Type: 748

Typical specifications for ±15-volt supply:

Parameter	Value	Unit	Limit
Gain	200,000		(50,000 minimum)
Input Offset Voltage	1.0	mV	(5.0 mV maximum)
Input Offset Current	20	nA	(200 nA maximum)
Input Bias Current	80	nA	(500 nA maximum)
Output Resistance	75	ohms	
Common-Mode Rejection	90	dB	(70 dB minimum)
Slew Rate	0.5	V/μs	

Remarks:

1. Can be compensated for unity gain with a single, external 30-pF capacitor.
2. Output short-circuit protected.
3. Differential input voltage: ±30 volts maximum.
4. Same pin configuration as 741.

CONNECTION DIAGRAM

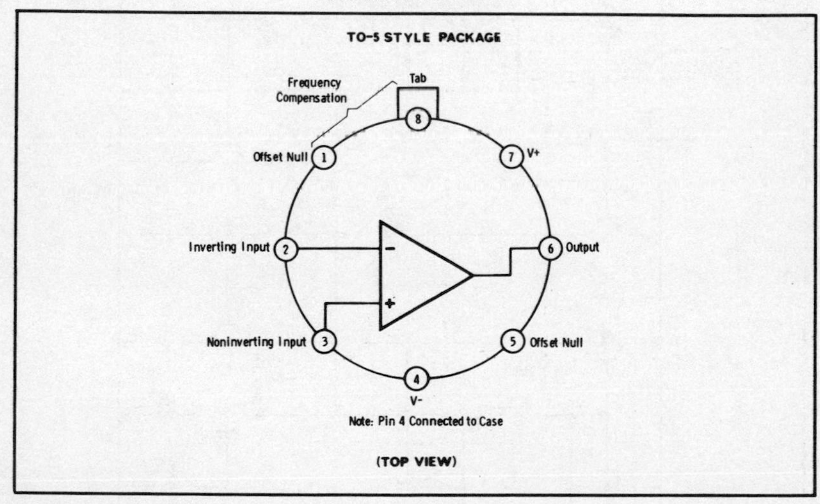

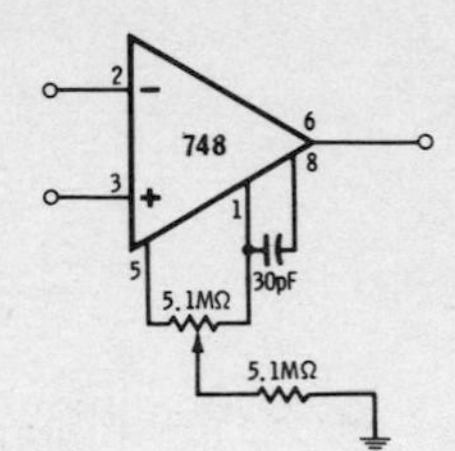

FREQUENCY COMPENSATION AND OPTIONAL OFFSET NULL CIRCUIT

Requires a single capacitor across the frequency compensation terminals. Use the smallest capacitor, 30 pF or less, which will provide stable operation. For fully compensated frequency response, see the 741 specification sheet.

SCHEMATIC DIAGRAM

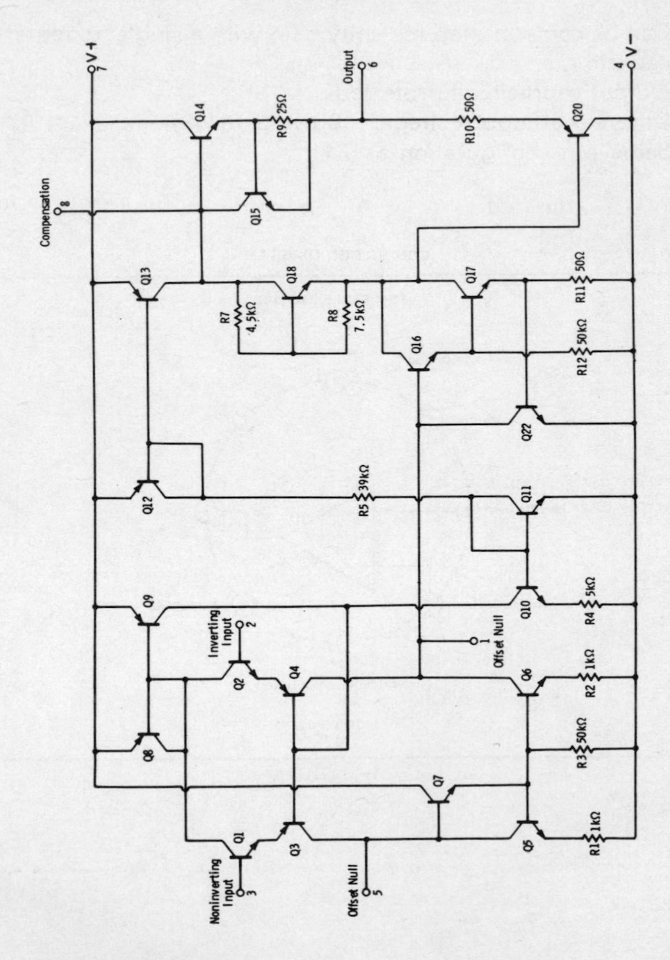

Type: 3130

Typical specifications for ±7.5-volt supply:

Gain	320,000	
Input Offset Voltage	8	mV (15 mV maximum)
Input Offset Current	0.0005	nA (0.03 nA maximum)
Input Bias Current	0.005	nA (0.05 nA maximum)
Common-Mode Rejection	90	dB (70 dB minimum)
Slew Rate	10	V/μs

Remarks:

1. MOSFET inputs for very low input bias current.
2. CMOS output stage allows output voltage to swing within a few millivolts of the supply voltage.
3. Maximum supply voltage limited to 16 volts (±8 volts).
4. The 3130 is uncompensated. Also available internally compensated (type 3160).

CONNECTION DIAGRAM

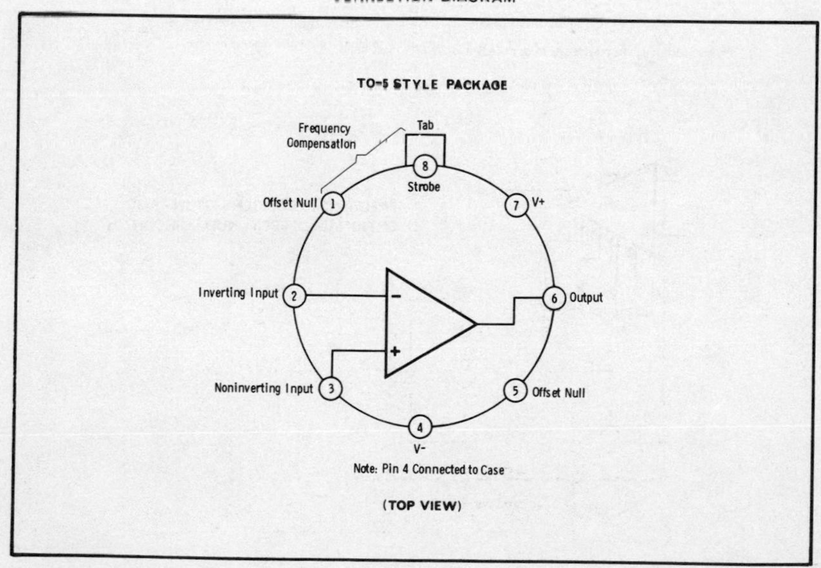

SCHEMATIC DIAGRAM

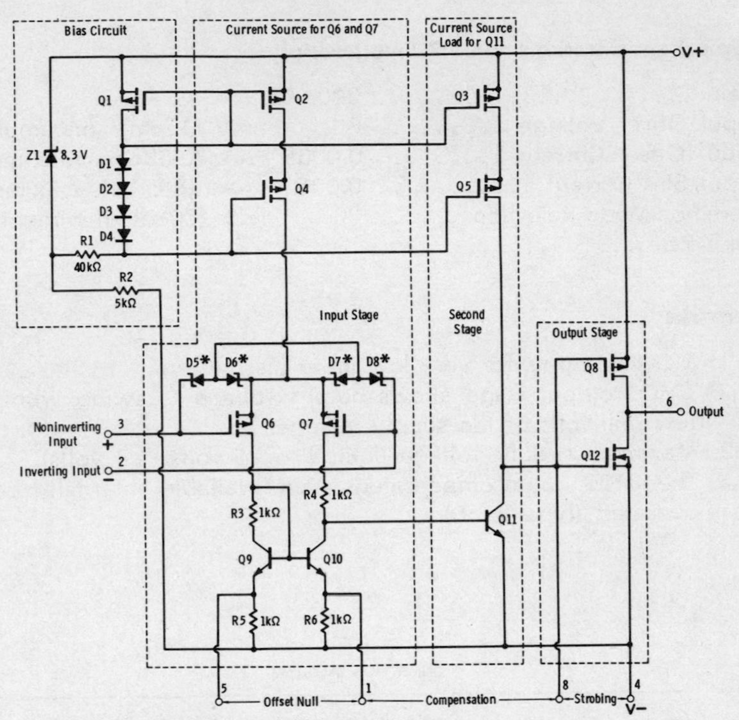

*Diodes D5 through D8 provide gate-oxide protection for MOSFET input stage.

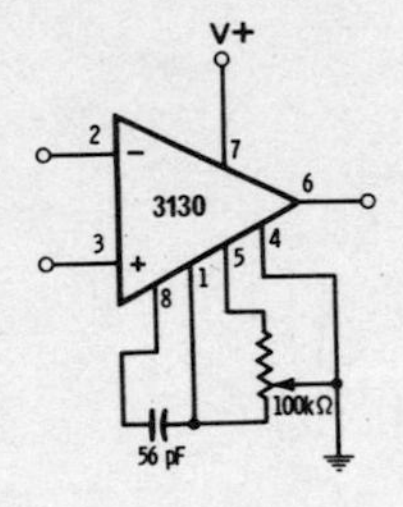

FREQUENCY COMPENSATION AND OPTIONAL OFFSET NULL CIRCUIT

Type: 3140

Typical specifications for ±15-volt supply:

Gain	100,000	
Input Offset Voltage	8	mV (15 mV maximum)
Input Offset Current	0.0005	nA (0.03 nA maximum)
Input Bias Current	0.01	nA (0.05 nA maximum)
Common-Mode Rejection	90	dB (70 dB minimum)
Slew Rate	9	V/μS

Remarks:

1. MOSFET input stage giving characteristics similar to those of the 3130.
2. Internally compensated.
3. Maximum supply voltage can be 36 volts (± 18 volts).
4. Minimum supply voltage is 4 volts (± 2 volts).
5. Same pin configuration as 741.
6. Output short-circuit protected.

CONNECTION DIAGRAMS

TYPE 3140

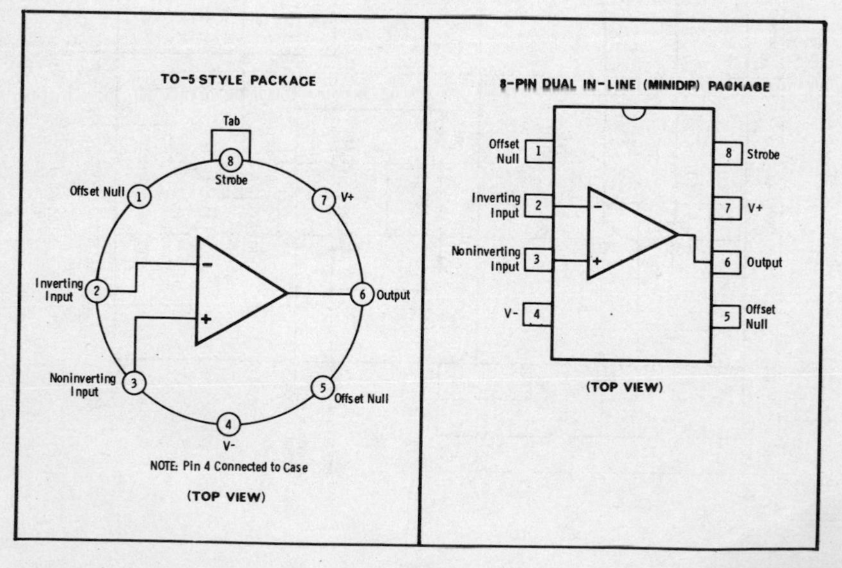

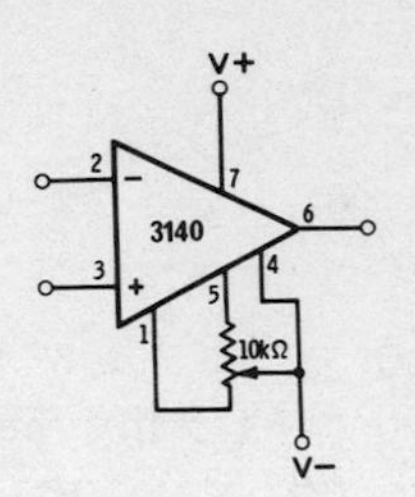

OPTIONAL OFFSET NULL CIRCUIT

SCHEMATIC DIAGRAM

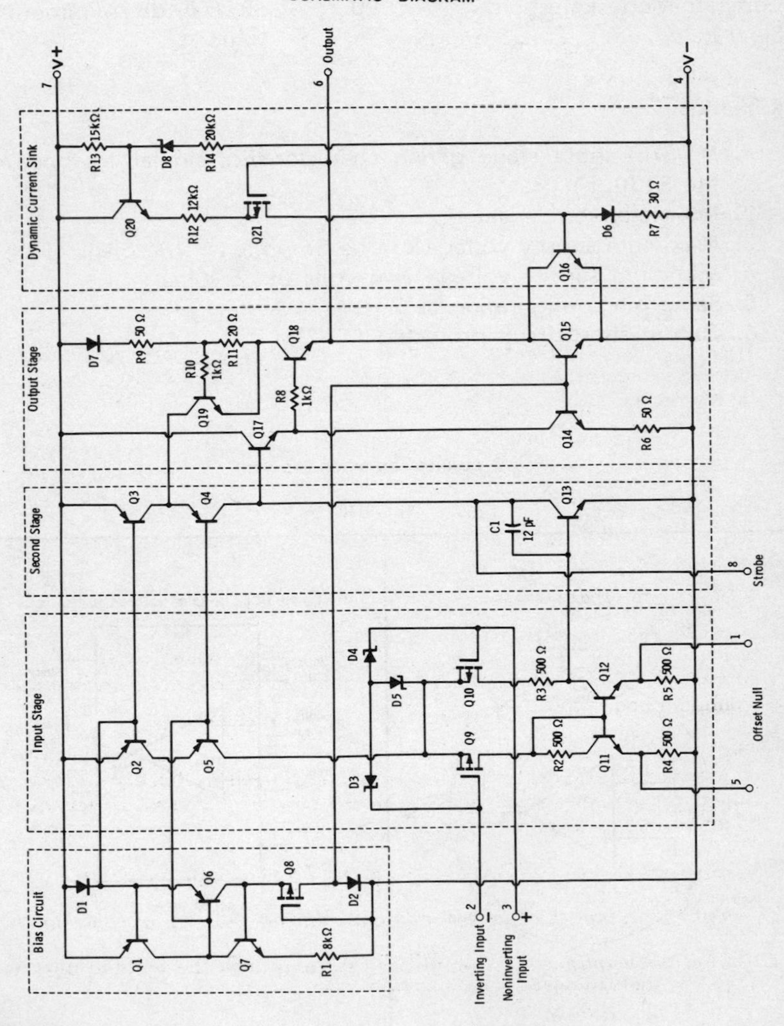

Glossary

BIFET—BIpolar-compatible, Field-Effect Transistor.

BIMOS—BIpolar-compatible, Metal-Oxide Semiconductor.

Carriers—The type of charge carrying a current. In a semiconductor, the carriers are electrons and holes. Electrons are negatively charged, and holes act as if they were positively charged.

Common-Mode Gain—The ratio of the output voltage of a differential amplifier to the common-mode input voltage. The common-mode gain of an ideal differential amplifier is zero.

Common-Mode Input—An input voltage common to the two inputs of a differential amplifier.

Common-Mode Rejection in Decibels—Twenty times the log of the common-mode rejection ratio.

Common-Mode Rejection Ratio—The ratio of differential-mode gain to common-mode gain.

Compensation—The shaping of the op-amp frequency response in order to achieve stable operation in a particular circuit. Some op amps are internally compensated while others require external compensation components in some circuits.

CMOS—Complementary-symmetry, Metal-Oxide Semiconductor.

Current Clamping—The output-current limiting feature of some op amps.

Differential Amplifier—An amplifier that amplifies the voltage difference between its two inputs.

Differential-Mode Gain—The ratio of the output voltage of a differential amplifier to the differential-mode input voltage.

Differential-Mode Input—The voltage difference between the two inputs of a differential amplifier.

Diffusion Current—The flow of a particular type of carrier in a semiconductor due to a concentration difference in that type of carrier. Carriers will flow from an area of high concentration to an area of low concentration.

Drift Current—The flow of carriers in a semiconductor due to an electric field. In the same electric field, holes and electrons will flow in opposite directions due to their opposite charge.

Epitaxial Layer—A single-crystal p-type or n-type material deposited on the surface of a substrate.

FDNR—Frequency Dependent Negative Resistance. A circuit used in some low-pass active filters.

FET—Field-Effect Transistor.

Frequency Compensation—See *Compensation.*

Gyrator—A circuit used in some active filters to simulate inductance.

Input Bias Current—The current that must be supplied to each input of an IC op amp to assure proper biasing of the differential input stage transistors. In specification sheets, this term refers to the average of the two input bias currents.

Input Offset Current—The difference between the input bias currents flowing into each input of an IC op amp, when the output of the op amp is at zero volts.

Input Offset Voltage—The voltage that must be applied across the two inputs of an op amp in order to produce zero voltage at the output.

JFET—Junction Field-Effect Transistor.

Latch-Up—The characteristic of some op amps to remain in positive or negative saturation after their maximum differential input voltage is exceeded.

Monolithic Integrated Circuit—An integrated circuit fabricated on a single piece of silicon.

MOSFET—Metal-Oxide Semiconductor Field-Effect Transistor.

N-Type Semiconductor—An area of a semiconductor with more electrons than holes.

Operational Amplifier—A high-gain differential amplifier.

Output Offset Voltage—The output voltage of a negative-feedback op-amp circuit when the input voltage to the circuit is zero. An ideal op amp has a zero output offset voltage.

P-Type Semiconductor—An area of a semiconductor with more holes than electrons.

Slew Rate—The maximum rate of change of the output voltage of an op amp as it swings from positive to negative saturation, or vice versa, in response to a square-wave differential-mode input.

Substrate—The supporting material on or within which the components of an integrated circuit are fabricated. The substrate for a typical monolithic fabrication is p-type silicon.

Index